CONTEMPORARY ETHICAL ISSUES

CONTEMPORARY ETHICAL ISSUES

ALBERT G. PARKIS
EDITOR

Nova Science Publishers, Inc.
New York

Copyright © 2006 by Nova Science Publishers, Inc.

All rights reserved. No part of this book may be reproduced, stored in a retrieval system or transmitted in any form or by any means: electronic, electrostatic, magnetic, tape, mechanical photocopying, recording or otherwise without the written permission of the Publisher.

For permission to use material from this book please contact us:
Telephone 631-231-7269; Fax 631-231-8175
Web Site: http://www.novapublishers.com

NOTICE TO THE READER
The Publisher has taken reasonable care in the preparation of this book, but makes no expressed or implied warranty of any kind and assumes no responsibility for any errors or omissions. No liability is assumed for incidental or consequential damages in connection with or arising out of information contained in this book. The Publisher shall not be liable for any special, consequential, or exemplary damages resulting, in whole or in part, from the readers' use of, or reliance upon, this material.

This publication is designed to provide accurate and authoritative information with regard to the subject matter covered herein. It is sold with the clear understanding that the Publisher is not engaged in rendering legal or any other professional services. If legal or any other expert assistance is required, the services of a competent person should be sought. FROM A DECLARATION OF PARTICIPANTS JOINTLY ADOPTED BY A COMMITTEE OF THE AMERICAN BAR ASSOCIATION AND A COMMITTEE OF PUBLISHERS.

Library of Congress Cataloging-in-Publication Data
Contemporary Ethical Issues / Albert G. Parkis (editor).
    p. ; cm.
Includes bibliographical references and index.
ISBN 1-59454-536-7
1. Contemporary Ethical Issues
[DNLM: 1. WL 360 P9643 2005] I. Parkis, Albert G.
RC377.P762   2005
614.5'9834--dc22   2005002153

Published by Nova Science Publishers, Inc. ✤ New York

CONTENTS

Preface		vii
Chapter 1	Developing Ethical Corporate Cultures: Three Examples of Organisations with Similar Approaches to Developing Ethical Corporate Cultures *Michael W. Small*	1
Chapter 2	Bioethics and New Age *Victor Tambone*	21
Chapter 3	Consent and Sensibility *Alasdair Maclean*	39
Chapter 4	Human Xenotransplantation: An Immunological and Ethical Challenge *Y. T. Ghebremariam, S. A. Smith, J. B. Anderson, D. Kahn and G. J. Kotwal*	63
Chapter 5	New Findings on Early Embryo Research and their Ethical Relevance *Miguel Ruiz-Canela*	87
Chapter 6	Equality, Priority and Levelling Down *Nils Holtug*	97
Chapter 7	A 'Special' Context?: Identifying the Professional Values Associated with Teaching in Higher Education *Bruce Macfarlane and Roger Ottewill*	113

| Chapter 8 | Utilitarianism, Repugnant Pleasures and Moral Explanation
Hugh Upton | **129** |

Index **147**

In: Contemporary Ethical Issues
Editor: Albert G. Parkis, pp. 1-20

ISBN 1-59454-536-7
© 2006 Nova Science Publishers, Inc.

PREFACE

This book presents theoretical and applied issues including ethical theory, moral, social, political, and legal philosophy. Issues include: biology and medicine, business, education, environment, government, mass media, science, agriculture and food production, and religion.

The purpose of chapter 1 was to look at three organisations (HMAS Stirling, Fleet Base West; Police Academy, Western Australia Police Service; and Geographe Enterprises Pty Ltd., Maintenance and Engineering Services) and see how they promoted and developed an ethical corporate culture within their organisations. It became obvious that the task of developing and promoting an ethical corporate culture in each organisation was being addressed in a very professional manner in all three organisations. The Navy as part of the Australian Defence Force (ADF) has a set of values (*HHCIL*) which partly overlap with the more general values of the Department of Defence *(imPLICIT)*. The submarine service, an elite group within the Navy, has added one its own viz. *esprit de corps*. The WA Police Service has approached the task differently, prompted partly by findings of Royal Commissions into inculcating an ethical approach to the way they carry out their duties. A formalised approach including a dedicated unit within the Service, discussion groups, seminars and publications are all part of the Police Service's way of promoting an ethical culture. Geographe is a family owned and operated business. It relies more heavily on an indirect approach based on traditional family values. However, the consensus was that three conditions were essential for developing, maintaining and promoting an ethical corporate culture viz. (i) CEOs and senior executives were ultimately responsible for the ethicality of their organisations, (ii) formal training programs were necessary to impart the required knowledge, and (iii) formal mechanisms were essential to facilitate the reporting of any behaviour of organisational members that was deemed to be wrong, unethical or illegal.

The phenomenon of New Age can be hardly outlined. In fact it is difficult to give a concise explanation of it. Through the *via negationis* (the defining way that goes on through the enunciation of what the phenomenon to be explained "is not"), can be said that it is not a movement, it is not a religion, it is not a sociological or anthropological theory and it is not, in precise meaning, a new political system. But at the same time we can identify new age movements, a new age religious view, new age sociological and anthropological theories. The author of chapter 2 thinks that the first defining character of the NA is its alogic-subjectivist's constitution. In this regard the authors completely agree with Introvigne when he points out that the main and deeper phenomenon's root is the epistemological one, even if we think it is a new kind of emotivism (subjectivist, by definition) more than a subjectivist voluntarism.

Consent is too complex to be understood fully in any one theoretical model'.[1] In chapter 3 the approach is that a core concept may be isolated and it is the peripheral attributes that are the seat of most disagreement. The authors intend to take a fresh look at consent and to cut away the excess flesh, which should make it easier to distinguish those things directly required by consent from those that need independent justification. In the first part they will consider the concept of consent. In the second part they will examine the moral basis that justifies consent and in part three they will discuss the implications of these prior discussions for the attribution of legal responsibility. The topic is voluminous and so the discussion must necessarily be selective. Its primary aims will be to reveal hidden or underlying assumptions and to explore the relationship between consent, autonomy and the responsibility for outcome

Despite the preliminary success in cellular and tissue xenotransplantation, the transplantation of solid organs across species has been a monstrous challenge in the history of xenotransplantation research. In principle, the transplantation of organs from taxonomically related concordant species has a survival advantage over the use of organs from unrelated discordant species. Yet, clinically, the survival of concordant xenografts has only been measured in days. Moreover, the use of organs from non-human primates may not be feasible due to logistic and ethical barriers. However, to benefit from non-primate organs, the biochemical and metabolic rifts created by the phylogenetic differences need to be resolved. Different haematological and genetic manipulations have been tried alone or in combination to prolong graft survival. Convergence of these strategies together

---

[1] Alderson, P. Goodey, C. "Theories of consent" (1998) 317 *British Medical Journal* 1313 at 1315.

with the administration of soluble complement control proteins, such as the vaccinia virus complement control protein (VCP), may provide a useful adjunct to prolong the survival of xenografts. However, a number of ethical and moral issues have obscured the topic of xenotransplantation and introduced yet another challenging barrier. Chapter 4 will discuss the immunological and bio-psychosocial issues involved in human xenotransplantation.

Recent research on the human embryo is revealing us surprising insights on the biology of the earliest stages of our own life as reported in chapter 5. The description of the changes -- molecular, genetic or structural -- taking place from fertilization to blastocyst formation makes a story that fills us with wonder and amazement. The novelty is twofold. First, it reveals us the untold variety, exactness and fine integration of the mechanisms and processes involved, and replaces the old crude sketch of spherical zygotes and identical blastomeres with a lively and elaborate depiction of dissimilar cells endowed with axis and poles, faces and asymmetries. Second, the new biological findings bring with them some factual data that add fresh impetus and deeper understanding to the enduring controversy on the ethical status of the early human embryo.

In chapter 6, the author first considers John Broome's account of the difference between egalitarianism and prioritarianism. He suggests that while his account plausibly stresses the relational aspect of equality, it does not quite capture its relational nature. Instead, the author offers a sketch of an alternative account or, more precisely, he points to some core intuitions about equality he believes that any plausible account of egalitarianism should accommodate. On the basis of this minimal account, h then considers (and criticizes) Marc Fleurbaey's claim that there really is no distinction to be made or, more precisely, that prioritarianism is really just a version of egalitarianism

Chapter 7 draws on the philosophy of higher education and existing codes of professional values as a basis for analysing the distinctive ethical challenges of teaching in higher education. For those who teach in higher education there are many values that they share with colleagues in schools and colleges, including respect for learners, collegiality, scholarship and a commitment to reflective practice. Additionally, however, they face a number of ethical challenges that, to some extent, distinguish them from teachers in other settings. These include protecting the academic freedom of students stemming from the goal of promoting student criticality; ensuring respect for learners derived from the concept of adulthood and the principle of andragogy; and accommodating a series of 'dual' roles which define academic identity. While the character of these challenges may vary between countries, arguably they are of international concern in higher education forming a distinctive basis for the identification of universal, professional values.

In proposing that actions are made morally right by their production of something held to be good, utilitarians have traditionally had in mind what we can call a natural commodity or state. To call something 'natural' is not of course a philosophically straightforward assertion but the relevant implication of this for our purposes is that it can be identified and characterised without recourse to some prior moral evaluation. The starting point of chapter 8 will be this traditional conception of the theory, and the problem to be considered will be outlined initially in terms of just one of the examples, the production of pleasure. Whilst this choice among natural goods may strike some as tendentious, it is made only for ease of exposition. It will be argued later that the selection is actually immaterial with respect to the issue in question.

In: Contemporary Ethical Issues
Editor: Albert G. Parkis, pp. 1-20

ISBN 1-59454-536-7
© 2006 Nova Science Publishers, Inc.

Chapter 1

DEVELOPING ETHICAL CORPORATE CULTURES: THREE EXAMPLES OF ORGANISATIONS WITH SIMILAR APPROACHES TO DEVELOPING ETHICAL CORPORATE CULTURES

Michael W. Small
Curtin Business School
Bentley, Western Australia 6102
phone: (+ 61 8) 9266. 7751 e-mail: SmallM@cbs.curtin.edu.au

ABSTRACT

The purpose of this chapter was to look at three organisations (HMAS Stirling, Fleet Base West; Police Academy, Western Australia Police Service; and Geographe Enterprises Pty Ltd., Maintenance and Engineering Services) and see how they promoted and developed an ethical corporate culture within their organisations. It became obvious that the task of developing and promoting an ethical corporate culture in each organisation was being addressed in a very professional manner in all three organisations. The Navy as part of the Australian Defence Force (ADF) has a set of values (*HHCIL*) which partly overlap with the more general values of the Department of Defence *(imPLICIT)*. The submarine service, an elite group within the Navy, has added one its own viz. *esprit de corps*. The WA Police Service has approached the task differently, prompted partly by findings of Royal Commissions into inculcating an ethical approach to the way they carry out their duties. A formalised approach including a dedicated unit within the Service, discussion groups, seminars and publications are all part of the Police Service's way of promoting an ethical culture. Geographe is a family owned and operated business. It relies more heavily on an indirect approach based on traditional family values. However, the consensus was that three conditions were essential for developing, maintaining and promoting an ethical corporate culture viz. (i) CEOs and senior executives were ultimately responsible for the ethicality of their organisations, (ii) formal training programs were necessary to impart the required knowledge, and (iii) formal mechanisms were essential to facilitate the reporting of any behaviour of organisational members that was deemed to be wrong, unethical or illegal.

INTRODUCTION

The chapter consists of three sections. The first takes the position that ethical theories and organisation theories are complementary, and that one area supports and reinforces the other. 'Ethical theories' covers a wide range of specialisations such as deontology, teleology, utilitarianism, and theories of justice. 'Ethical' in the context of this paper is used to describe behaviour which is right or wrong, good or bad. 'Organisation theory' is concerned with explaining the origins, the functioning and the operations of organisations. Organisation theory and ethics are two separate fields yet both are involved in the process of developing ethical corporate cultures. The two areas can be seen as complementary when organisation values are analysed. In the present study the organisation's values are also moral values. For example, an ethical corporate culture assumes right and wrong behaviour in an organisational context, and so values such as justice, respect and open communication (transparency) become relevant. Tsoukas and Knudsen (2003: 27) and Nielsen cited in Tsoukas and Knudsen (*ibid.* 476 ff.) have argued that although OT and Business Ethics are separate disciplines, it is impossible to separate them.

Developing an ethical culture *in vacuo* i.e. without taking into account an organisation in which to locate it is pointless. Similarly, developing an organisation without taking into account an ethical dimension is mindless. Undertaking the task of developing an ethical corporate culture therefore requires an understanding of ethical theories in the broadest sense, and an understanding of the way organisations are structured and operate.

The purpose of this chapter therefore was to illustrate how ethical corporate cultures could be promoted and developed, and how an understanding of both organisational theory and ethical theory could facilitate this process. We know that within the 'nation-state' a political theory provides direction for the majority of its people. We also know that individuals have their own moral codes based on a particular moral theory (or perhaps no moral code at all) to provide direction through life.

Organisations including corporations in this country seem reluctant to proclaim an ethical or moral code of their own, although increasing numbers of business corporations are including (token) references to codes of conduct in their annual reports. They survive, prosper and deliver by exercising 'command and control', by the liberal use of 'power' and 'authority', and by supporting the idea of the 'chain of command'. Individuals who work in organisations clearly would prefer an environment which promotes harmony and goodwill. The issue therefore is to show that an ethical corporate culture can meet the needs of the individual, yet at the same time see that the goals of the organisation are realised.

The issues to be addressed here are complex. In developing an ethical corporate culture, ethical theory and organisation theory are dependent on and complementary to each other. Contemporary management theory recognises and is becoming more accepting of moral theory. Concepts such as the 'power/authority' nexus, the 'command/control' principle and the notion of 'compliance', with all their negative and unpleasant connotations have to be accepted as necessary to see that things get done.

At a personal level, the process of establishing meaning and purpose in an individual's work situation may counteract the negative aspects of corporate life. If an individual is given the opportunity of reaching his/her potential, of making a contribution and associates with good colleagues in an ethical environment then the chances of an ethical corporate culture developing are greatly improved.

At an organisational level, equality and fairness may not always be present. However certain procedures (work processes such as finishing tasks, behavioural processes such as decision making and communicating, and change processes in which individuals grow and adapt over time) can promote and foster the development of positive ethical corporate cultures.

Organisations have to deal at some time with ethical and moral concerns. These can be related to the leadership structure and the extent to which that structure is based on integrity. This area is recognised as an area of interest in organisation theory, and in the emerging area of philosophy of management. 'Moral responsibility', 'moral management, 'the role of the moral manager' and 'corporate social responsibility' are areas currently receiving attention. In the area of corporate social responsibility, Paine cited in Dubbink (2004: 26) distinguished between 'compliance strategy' and 'integrity strategy'. There are also economic, legal and philanthropic implications for organisations as well as the moral concerns, prompting questions such as 'does the manager operate within the letter and/or the spirit of the law? Issues such as inequality and fairness in the workplace and maintaining one's professional integrity are included. These will be referred to later in the chapter.

ETHICAL THEORY AND ORGANISATIONAL THEORY

The idea that idea that ethical theory and organisational theory are complementary and that one area can support and reinforce the other was introduced in the Introduction. Organisation theory and ethical theory, two separate and distinct fields (Tsoukas and Knudsen: *ibid.: 476),* can both be seen as having a stake in developing ethical corporate cultures. When any organisation begins the process of developing a policy, initiating an action or making a decision the organisation's wheels begin to turn, and organisation theory comes into play. The two criteria of effectiveness and efficiency were normally the principal criteria used in reviewing policies, actions or decisions. These have now been strengthened by the addition of additional criteria which have introduced the ethical/philosophical component into the policy-making process.

Justice

Justice is a major criterion. Questions to be addressed include: 'does the particular policy, action or decision distribute any benefits that may occur fairly?' Rawls' explanation of justice and distributive justice in particular would be appropriate here. Issues such as equity, fairness and impartiality are built into the argument. This criterion introduces the economic aspect into the argument by inferring that justice is the act of giving each person his due i.e. treating equals equally and unequals unequally.

Rights and Duties

Rights and the duties which are associated with rights introduce the question: 'does a particular policy, action or decision respect human rights?' Are processes such as free consent, privacy, conscience, free speech, and due process transparent and are they built into the system?

Utilitarian Theory

John Stuart Mill (1863) in his 'utilitarian theory or greatest happiness principle' claimed that behaviours that were moral produced the greatest good for the greatest number.

Caring

'Caring' is a fourth criterion. Does the modern organisation 'care' for its personnel? This idea of 'caring' featured recently in a report into a "redress of grievance" *The Bulletin,* (April 20 2004, p. 21). The report based on the Toohey Inquiry, was prepared by a naval barrister and solicitor into alleged injustices to an army intelligence officer.

Virtue Ethics

Finally, 'virtue ethics' (honesty, good personal relations, teamwork and trust) features in this approach in identifying ethical issues in the three organisations. The process of scrutinising the ethicality of an organisation can be extended by analysing the level of trust, the commitment, the degree of effort in an organisation, and the extent to which these are pursued. Does an organisation accept that it has a duty to recognise moral problems which may arise? Does an organisation apply moral reasoning in terms of determining what is right? And finally does an organisation have sufficient moral courage (integrity) to undertake particular courses of action? These areas by coincidence would seem to be the subject of a special issue of *Business Ethics Quarterly* to be produced in 2005. The subject for this special edition is "The Ethics of Organisational Ethics Initiatives".

FRAMEWORKS FOR ORGANISATIONAL ANALYSIS

Organisations can be analysed according to their goals, technology and structure. Perrow cited in Hodgkinson (1978: 28ff.) identified the purposeful aspect of organisations. He developed a fourfold classification viz. (i) output goals (goods and services), (ii) system goals (maintenance and growth), (iii) product goals (quality, quantity and demand), and (iv) derived goals which included political aims and employee development. An ethical dimension would fit into this category.

Blau and Scott *(ibid.)* classified organisations as mutual benefit associations, business concerns, service organisations, and commonweal organisations. Katz and Kahn *(ibid.)* described the productive, maintenance, adaptive, and managerial-political types of organisation. Mooney cited in March and Simon (1967: 30), Hodgkinson (1978: 10) and in the earlier work of Gulick and Urwick (1937) described organisations in terms of their perpendicular coordination, horizontal coordination, leadership, delegation and authority. Homans cited in Litterer (1967: 120 ff.) argued that there are three elements of a social system which he classified as follows:

Activities

Activities are the things people do. Whether as required by the external system and part of the job or whether they emerge out of the internal system such socialising on or off the job.

Interactions

Interaction occurs when two people come together in some way so that one person has an effect on the other. It is similar to the activities described above.

Sentiments

Sentiments are concerned with the internal states of a person such as his/her emotions or feelings or beliefs or values. They can also include prejudices, ideals and hopes.

Values are ideals or standards. Norms are more specific and can be measured. Litterer's *(ibid.:*124) ideas could so easily provide the model for detailed analyses of the three organisations in this study. Simon (1965: 20-21) referred to the principles of specialisation viz. a hierarchy of authority, a limiting span of control, and grouping the workers according to purpose, process, clientele or place.

Finally, Hatch (1997) provides a framework which could accommodate the three organisations under review viz. the environment, their strategy and goals, the technology they utilise, their social structure and culture, and the physical structure of their organisation. Any of these criteria could be used to provide the theoretical framework for an investigation into the extent to which an ethical corporate culture was being developed.

HOW ETHICAL CORPORATE CULTURES DEVELOP

This section addresses the way ethical corporate cultures develop, and how an understanding of organisation theory and ethical theories can facilitate this process. Organisations and this includes corporations do not seem to proclaim an ethical or moral code of their own, although increasing numbers of corporations are now beginning to include brief references to codes of conduct in their annual reports. Organisations survive, prosper and deliver by using a number of strategies e.g. the command and control principle, the active use of power and authority, and by supporting the chain of command. The problem therefore is to show how a corporate culture can take into account the needs of the individual, but at the same time see that the goals of the organisation are met.

The ideas presented here show that in developing an ethical corporate culture, ethical theory and organisation theory can be seen as dependent on and complementary to each other. However, this line of thinking is made more complex by recognising that a hierarchy, the authority which goes with it and the associated subordination are necessary if organisations are to complete their tasks.

The implication here is that equality, a sense of fairness or justice and a moral approach to organisational issues are not automatically guaranteed. Some top level managers have demonstrated that all is not well with business morality. Hence the current interest in contemporary management theory becoming more inclusive of moral philosophy. Managers, including those at the highest level, should take lessons in a variety of subjects (humility has been suggested) and to conduct themselves and their businesses in more appropriate ways. Arrogance, ignorance and a disregard of the law are no longer acceptable requirements for today's (business) managers. In fact the exact opposite of these three characteristics is now seen as essential. High profile chief executives have been charged with fraud, conspiracy to commit securities fraud, insider trading, various corporate scandals and suspect accounting practices that have led to the collapse of their corporations. In one recent case (March 2004) a business woman in the United States was convicted on 'obstruction of justice' charges. However, the conviction seems to have had little or no impact on her designer furniture business. Her business manager reported (23 April 2004) that 'sales were up, and everything was just fine'.

At the same time the power/authority nexus, the chain of command/control and the notion of compliance, all with their negative and unpleasant connotations have to be accepted as necessary evils to see that organisational goals are met. Establishing meaning and purpose in an individual's work situation may help to counteract the negative aspects of corporate life at a personal level. If an individual is given the opportunity of reaching his/her potential, makes a contribution, and associates with good colleagues in an ethical and collegial environment then the chances of an ethical corporate culture developing are greatly improved. At the organisational level, equality and justice or fairness are not always present. However if certain procedures referred to earlier (see p.3) are promoted the development of a positive ethical corporate culture can be accelerated.

Reference is made in the following section to six criteria which could assist in determining the extent to which the development of an ethical corporate culture would be likely. Most are well-known and present no surprises. These criteria have all been utilised in the three organisations under review

Code of Conduct

The first criterion is a visible, explicit code of conduct or code of behaviour that is known and followed by the entire workforce. As an example, the members of a rugby league football club were severely criticised in the press and fined by their club because they had ignored the club's code of ethics in respect to appropriate dress behaviour. Some members of the team who were 'assisting police with their inquiries' had to attend a court hearing into their behaviour. They arrived wearing very casual and inappropriate clothing. In the three organisations in this study, the code of ethics criterion is recognised as an aspect of good management practice.

Ethics Training

Ethics training for managers, the second criterion, is also recognised as good management practice, and is a regular feature of police training. The WA Police Academy is endeavouring to promote an 'integrity strategy' as opposed to a 'compliance strategy' referred to earlier (Paine, *ibid.*: 26). Most organisations now include some reference to ethics training programs in their annual reports. This may refer to short and/or half-day executive courses or attendance at full day seminars presented by visiting management gurus who are specialists in particular areas.

Internal Programs

Internal programs to resolve ethical conflicts have been set up by Australian police services. The RAN has a different approach and makes use of the Divisional system. Again the Navy takes the approach of promoting the 'integrity strategy' as opposed to the 'compliance strategy'. These will be referred to later in the paper. Geographe, the engineering firm, is different. Their approach is more informal and pre-emptive. They believe that this approach can forestall any untoward occurrences.

Ethics Review Committees

Ethics review committees can be used to determine if ethical procedures are being followed and if there is a need to focus on particular aspects.

Rewards and Punishments

All three organisations agree that a system of appropriate rewards and punishments should be in place.

Attitude of Top Management

Finally the attitude of senior management is highly significant in determining the extent to which an ethical corporate culture is wanted, promoted and develops. Gilmartin (2003: 9) speaking at Bentley College addressed the topic of what constitutes a strong ethical culture. He identified three important goals if a company were to prove its commitment to the highest standards of ethics and integrity viz.:

I) the organisation's top leaders must set the right tone.
II) the organisation must offer formal training in ethics and standards of conduct.

III) the organisation must provide formal mechanisms, both inside and outside one's organisational structure, for reporting any wrongdoing, should it occur or be suspected of occurring.

AN OVERVIEW OF THE THREE ORGANISATIONS

Two of the organisations, HMAS Stirling and the Police Academy, are similar in many respects. They are self-contained units within much larger organisations. HMAS Stirling, Fleet Base West is the submarine base and a shore establishment of the Royal Australian Navy. The second is the Police Academy, Western Australia Police Service. The third organisation, Geograph, is a family owned and operated business specialising in maintenance and engineering services. It is based in Bunbury, with branches in metropolitan Perth and the larger country centres. This organisation is included because it represents the type of organisation more usually included in a business research undertaking.

The WA Police Academy and HMAS Stirling have many are similarities. Both services are hierarchical, with clearly defined chain(s) of command and authority. Both of these organisations rely on subordination (although a high degree of mutual respect was noticeable on the part of all ranks in both of these services) to achieve their goals. Both services have had a bad press recently. A number of fatalities in the RAN, and the findings of a Royal Commission into the WA Police Service (2 March 2004) have given the public an unfortunate impression of them, an impression which is not altogether warranted.

HMAS Stirling, the submarine base of the RAN provides a pragmatic approach to organisational structure in terms of areas of responsibility and chain(s) of command. To illustrate, COMSUBGRP i.e. the Commodore of the submarine flotilla reports to the Maritime Commander in Sydney who in turn reports to the Chief of Navy in Canberra. COMSUBGRP is also SNOWA (i.e. Senior Naval Officer Western Australia), notwithstanding that there is a Commanding Officer (naval rank Commander) of HMAS Stirling who reports through a different line to the Naval Systems Commander. The Commanding Officer's role and responsibilities are more related to the administrative and functional side of the base. The roles and responsibilities of SNOWA and the Commanding Officer could seem to be at odds, but the fact that arrangement works says a great deal about the common sense approach that prevails in the Navy in handling these issues.

The organisational structure and chain(s) of command within the base are areas of interest. The base Commanding Officer has administrative and disciplinary responsibilities for officers who may be senior to him by one or two ranks in the Naval hierarchy. These could be officers responsible for lodger units at the base, which have a different chain of command for operational matters.

How this arrangement works (which does seem contrary to business and other organisations with strictly defined lines of authority) and the degree to which it is successful could provide the basis for a more detailed study at a later date.

HMAS Stirling also illustrates a very strong sense of social responsibility and social awareness in respect to the ecology and the natural environment in which the base is situated. To readers who are unfamiliar with the fauna and flora of Western Australia, the bush surrounding the naval base is the last remaining natural habitat of the tamar (a miniature or scaled down kangaroo). It is a nocturnal marsupial which emerges at dusk, hops about and shows no fear of the uniformed personnel. The surrounding bush is also home to the venomous tiger snake. As the Commanding Officer said, HMAS Stirling has very little problem with unauthorised entry to the base.

Geograph is a medium sized manufacturing engineering firm employing approximately one hundred and sixty people. It specialises in the supply of spare parts for the mining industry, the re-engineering of components and undertakes analyses into the failure of high strength alloy materials. In other words it provides the answers to why engineered parts break. Geograph has branches in Bunbury, Perth, Kalgoorlie, Port Hedland, Townsville and Broken Hill. According to the customer service director, Geograph is firmly committed to the principle of 'corporate social responsibility (CSR)' in all of its operations, and employees of the company are aware of this aspect of company policy. Dubbink (2004: 26) defined CSR as follows:

> CSR can roughly be defined as the opinion that market organizations have a certain responsibility when it comes to public issues, a responsibility that goes beyond the demands of law and common decency. I have tried to arrive at a more precise definition by bringing CSR into line with standard liberal-democratic notions about the rights and obligations of actors. As a consequence, this precise definition can best be described as ideal-typical.

DEVELOPING AN ETHICAL CORPORATE CULTURE: POLICE ACADEMY WA POLICE SERVICE

Recent media coverage concerning the conduct of police officers (involving those who were associated with the illicit drug scene) presents a disturbing picture. 'Four Corners', an ABC TV documentary program, reported the findings of an investigation (8 March 2004) which was undertaken into the Victoria Police. The investigation focused on a number of police officers who had systematically undertaken unauthorised transactions of illicit drug related material. These actions involved of very large sums of money and involvement with members of the illicit drug underworld.

A more recent media news item (10 March 2004) reported that the head of the Victoria Police drug squad had been arrested and would face criminal charges. In this expose of the Victoria Police, the ABC TV reported (24 May 2004) another more personal story of one particular police officer who had blown the whistle on corrupt senior police. This police officer shown in the ABC TV program (24 May 2004) had taken a different role. He had decided to pursue corrupt members of the Victoria Police and bring them to justice. The story takes a turn for the worse with the report that two people who were described as 'informers', and likely to be major prosecution witnesses in a forthcoming trial of corrupt police were murdered in their home.

In Western Australia, the Final Report of the *Royal Commission into Whether There had Been Any Corruption or Criminal Conduct by Western Australian Police Officers* was released (2 March 2004). Both of these examples tell very much the same story. A number of police officers became involved with dealers in illicit drugs. These police officers were subsequently compromised by their association with the illicit drug dealers. The details may be different, but the end results were the same. Officers in two State police services were compromised and corrupted, and so both Services became vulnerable to outside pressures.

While corruption within an organisation is one area of interest in this study, the number of personal tragedies which are known and recorded should not be ignored. The writer was involved with a group of senior commissioned police officers undertaking executive training programs. One member of this group was a chief superintendent in his mid-thirties. He was being spoken of by his peers as a possible commissioner at some future time. This person later appeared in a TV documentary where he was clearly identified receiving money from known members of the drug underworld. The question has to be asked how do police fall into this situation, despite the warnings and the advice from senior officers, and the training they received from their instructors at the police academies.

The Personal Level

At the micro or personal level, new recruits at the WA Police Academy are introduced to a range of strategies, which are intended to alert them to situations where unethical and improper professional behaviour may occur. Again 'integrity strategy' is the motivation behind this approach as opposed to 'compliance strategy'.

During the initial recruit training period, each new police constable undergoing training receives a hip-pocket sized bound and embossed copy of the *WA Police Service Code of Conduct*. This personal document introduces the new recruit to values such as *ethics, integrity and professional conduct*. It highlights in red six non-negotiable principles of conduct: e.g. *honesty, respect, fairness, empathy, openness, and accountability*. The document identifies a range of potential problem areas which the new recruit will meet in his/her career e.g. *conflicts of interest, problems associated with receiving gifts or benefits, discrimination, inappropriate communication, and problems associated with drugs/alcohol*. One section deals with methods of reporting incidents which involve *serious misconduct and corruption, unprofessional conduct or unethical behaviour*. This is done by reporting the incident to a senior officer, the ombudsman, or by using a confidential phone line. The 'Blue Line' is a confidential, no trace and dedicated telephone line for personnel who are concerned about the conduct of other members of the WAPS. In addition to the above mentioned 'The Supported Internal Witness Program' (SIWP) exists to provide further support and assistance to anyone who has taken the difficult step of reporting corruption or serious misconduct. The last page of the document contains a page requiring the new recruit's signature. It is then counter signed by his/her supervisor. This is evidence or proof that the person who owns the booklet has indeed read and studied it. The page is then removed, sent to HQ and kept on file. If at some future date the officer transgresses and is guilty of unprofessional conduct, the signed page is retrieved and shown to the officer for comment.

A second pocket-size document, *Charter of Rights*, sets out the WA Police Service's position in respect to the Authority of Police, Community of Rights and Community Obligations.

A third document is more substantial. It addresses the subject *ETHICS RECRUIT LEVEL*. It is professionally bound and presented, and the content matter is accurate and well set out. This document contains the Code of Ethics, Section 10 Police Act, the assessment procedure, and topics for a later essay examination. The document also contains information on the seven virtues of police integrity viz. *prudence, trust, effacement of self-interest, courage, honesty, justice, values,* factors that impact on moral decision making, tools for making the right decision, the service philosophy, and a more lengthy section on the Police Act 1892 and Police Force Regulations 1979.

Part IV Section 402 of the Regulations refers to Provisions relating to behaviour, and Section 608 refers to members not compromising themselves by receiving bribes, presents or placing themselves under any pecuniary obligation. The volume includes sections highlighting key ethical issues e.g. *ethical guidelines, ethical dilemmas, work sheets* on topics chosen for class discussion, and a final section which includes realistic topics such as '*accountability and honesty*', '*diversity and discrimination*', and '*supervision and accountability*'.

It could be assumed therefore that if the new recruit had read and understood the contents of his/her two pocket size books and the substantial bound volume then s/he should have very little problem with any future moral issue or ethical dilemma.

The Organisational Level

At a macro or organisational level the strategies put in place to develop an ethical corporate culture are equally impressive. The 'Standards Development Unit' was established in 1996, and reported to an inspector. It was established to:

> enhance the standard of ethical performance within the Western Australia Police Service.

The unit was given the responsibility of contributing towards a change in policy, and of developing strategies for the continuous improvement of ethics and integrity. The unit also provided support for a range of other activities associated with raising the awareness of ethical behaviour.

In 2002 the unit was re-constituted, upgraded and renamed the 'Professional Standards Portfolio' reporting to an assistant commissioner. The structure (July 2004) now has four streams. These are as follows:

(i) Ethics and Standards Division (Superintendent-in-charge). Within this Division is the Standards Development Unit (Inspector-in-charge) who reports to the Superintendent-in-charge of the Ethics and Standards Division. (ii) Complaints Administration Centre (Superintendent-in-charge). (iii) Internal Affairs Unit (Superintendent-in-charge). (iv) Management Audit Unit (Manager-in-charge).

The Strategic Plan

In the Strategic Plan 2001-2006, the Minister for Police and Emergency Services wrote:

> The Police Service has been granted special powers by the community and in return police officers are expected to set the highest standards of ethics, integrity and professionalism.

The Commissioner addresses, *inter alia,* 'the business of policing in the future, but also the type of organisation we seek to be'. The Strategic Plan includes a section on 'Vision, Mission and Values', and another where the values are clearly defined. These concepts keep re-appearing in different publications, so it could be assumed that the recruit and the serving officer would be familiar with them and would know exactly what was expected of them in carrying out their professional duties.

DEVELOPING AN ETHICAL CORPORATE CULTURE: GEOGRAPHE ENTERPRISES PTY LTD

'Geographe' is different in structure, technology, culture and hierarchy to both the Police Academy and HMAS Stirling. It is a family business specialising in maintenance and engineering services. According to the Customer Service Director, Geograph prides itself on its customer relations and its moral approach to correct and ethical business behaviour. As the Customer Service Director has emphasised more than once, Geograph was in business for the long haul. If Geographe's senior management wished to maintain their customer base they would have to keep treating their customers in a fair and reasonable manner. The profit motive was handled in away that was acceptable, and no accusations of excessive charges could be sustained. As a family operated business, the ethical guidelines, ethical code of practice and the policies which were in print form for the Police Academy and HMAS Stirling were not a part of Geograph's inventory.

There was an inbuilt assumption that some things, such as specifying an ethical approach to business did not need to be spelt out. The culture of the organisation was based on the family's values, and the expectation that any business transaction would be within both the letter and the spirit of the law. Geographe, in the words of the Customer Service Director, has an inbuilt ethical culture, which was woven into every policy. This culture also facilitated the detection of poor behaviour before it had time to become established. As a final check, even though the firm was a family concern, and therefore not required by law to undergo an audit, Geographe had elected to do this to determine that all the necessary processes were being carried out.

DEVELOPING AN ETHICAL CORPORATE CULTURE: HMAS STIRLING FLEET BASE WEST

The Department of Defence and the Navy as part of the Australian Defence Force (ADF) have an on-going and serious relationship with the study of ethics and justice in a military context. This became apparent as the research for this paper progressed. For example, the Department of Defence has in its collection a number of works by writers such as Brown and Collins (eds.) (1981), Ingram and Parks (2002), Preston (1996), Toner (1995) and Torrance (1998) (see reference list for full details). In addition the Department of Defence produces a number of high quality pamphlets aimed at informing/advising members of the Defence establishment about a wide range of ethical issues.

These are grouped into three categories: (i) DEFAC (Defence Ethics and Fraud Awareness Campaign) pamphlets. These cover subjects such as fraud and fraud investigations and personal rights. (ii) The Defence Ethics and Fraud Awareness Campaign (DEFAC) published in the DEAR (Defence Ethics and Resources) series. This is a quarterly newsletter examining ethical issues in the management of Defence resources. Subjects such as: 'A Study of Values and Ethics in Managing Resources in Defence', 'Why all the fuss about ethics?', 'The Problem of Chain Letters', 'The Whistle Blower-The Ethical Dilemma', 'Codes of Conduct', "What has ethics got to do with Defence?' are described and analysed in some detail. (iii) 'ethics matters' is a newsletter in which ethical issues in the management and use of Defence resources are examined. The subjects covered in 'ethics matters' are of interest in the context of this paper. Subjects such as the use of Defence telephone and computer resources, the ethical management of intellectual property, conflicts of interest, maintaining a values-based and ethical Defence Organisation, the misuse of government credit cards, and the inappropriate use of frequent flyer points are covered in this series. This publication is produced under the auspices of the Inspector General and issued at intervals of three to four months.

Fraud Control Plan No. 5 February 2004 was a paper prepared specifically for the Submarine Force Element Group (SMFEG). The purpose of this paper was 'to explain the strategy for fraud control in SMFEG, allocate responsibilities for tasks under the plan, provide leadership and direction, and describe the fraud control-reporting regime'. Areas reviewed in the paper included the management of consumables, official hospitality and gifts, international and domestic travel, allowances and purchasing. The paper includes a section 1.6 'Fraud Control Policy' which described fraud as having, *inter alia*, 'a debilitating effect on any organisation'.

Other papers in the collection included Torrance's (1998) explanation of 'natural law' and a discussion of Pope Paul VI's encyclical *Humanae Vitae* (1968). Torrance (*ibid.*) referred to well known scholars such as Alasdair MacIntyre, John Rawls and Robert Nozick. Torrance wrote (*ibid.*) that 'he wanted to apply some of these extended analogies to the Army, thinking of it also as a moral community at a time of change'. Brown and Collins' (Eds.) collection of essays was perhaps the most interesting work referred to in the study. The topics 'Moral and Ethical Foundations of Military Professionalism', and 'Ethics in the Military Profession: The Continuing Tension' illustrate the type of material that is available to Department of Defence personnel.

The Department of Defence as the overriding co-ordinating government department charged with managing the nation's defence assets uses the mnemonic *implicit* to illustrate the values required of its senior leadership. This becomes more meaningful if written as *imPLICIT* ('P'/professionalism, 'L'/loyalty, 'I'/innovation, 'C'/courage, 'I'/integrity, and 'T'/teamwork").

The Chief of Navy's document 'Our Mission 2001-2002' states:

> Our values guide how we will behave, how we will treat each other, what is important, and what bonds us together. Values are our source of strength; they are the source of moral courage to take action.

The Navy while recognising the Department of Defence's values has developed a set of its own based on the letters *HHCIL*. There is clearly some overlap between the two agencies. *HHCIL* can be translated as follows. 'H'/honour is the fundamental value on which the Navy's and each individual's reputation depends. The second 'H'/honesty stands for always being truthful and knowing and doing what is right. 'C'/courage refers to strength of character and doing what is right in the face of personal adversity, danger or threat. 'I'/integrity refers to a display of truth, honesty and fairness that gains respect and trust from others. 'L'/loyalty is being committed to each other and to one's duty of service to Australia. The submarine service has added one of its own to the Navy list viz. *esprit de corps*. While not strictly speaking a value, *esprit de corps* does seem to be appropriate for the special role that the submarine service plays in the defence of Australia.

The Divisional system which has been in place for many years in Royal Navies has been likened to the house or form master system. It was designed to protect the rights of the individual. In the Navy's words:

> the divisional system is designed to provide the appropriate framework for command and leadership across all ranks, and to enable fair and due process of personnel management issues. Each member of the divisional chain of command is required to maintain a values-based approach to both professional and personal behaviour and ethics.

In a large warship in earlier days, the complement was divided into divisions as follows: seamen, engine-room artificers, torpedo-men and young sailors. Each division was the responsibility of a divisional officer who was always available for advice. The ship's chaplain(s) often stepped into this role if and when needed. The system in today's RAN has obviously been modified, but the Divisional system remains.

CONCLUSION

The purpose of this chapter was to review the way three organisations (HMAS Stirling, Fleet Base West; Police Academy, Western Australia Police Service; and Geographe an engineering and maintenance firm) established and maintained an ethical corporate culture. It was argued that organisation theory and ethical theory were complementary, and that both areas could facilitate understanding in developing an ethical corporate culture.

In brief, all three organisations dealt with the task in a professional and systematic way. The Navy using a less direct approach recognised that there were issues that needed attention and had developed strategies to rectify these. The Department of Defence as the coordinating government department publishes a range of pamphlets to inform and advise members of the Defence establishment of their responsibilities. The Department of Defence, the Navy and the submarine service have identified a range of values which are introduced to all personnel by an indirect approach. This is in contrast to the Police Academy which has established a very formal approach with policies and procedures ensuring that every police officer is aware of his/her ethical responsibilities. Geographe is a family owned and operated firm. It relies on promoting family values i.e. an indirect approach to instil ethical values into their work force. In summary, it can be stated that all three organisations adhere to the belief that there are three major conditions that organisations must meet if they are to develop an ethical corporate culture viz.

I) the organisation's senior staff must set the right tone.
II) the organisation must offer formal training in ethics and in standards of conduct.
III) the organisation must provide formal mechanisms, both inside and outside the organisational structure, for reporting any wrongdoing, should it occur or be suspected of occurring.

REFERENCES

Brown, James and Michael J. Collins (eds.) 1981, Military Ethics and Professionalism A Collection of Essays, *The National Defense University National Security Essay Series 81-2*, National Defense University Press, Fort Lesley J. McNair, Washington, DC, 20319.

Dubbink, Wim. 2004, "The Fragile Structure of Free-Market Society", *Business Ethics Quarterly,* 14 (1), January.

Gilmartin, Raymond V. 2003 "Ethics and the Corporate Culture", *Raytheon Lectureship in Business Ethics*, Center For Business Ethics, Bentley College, Waltham, Mass. given by Raymond V. Gilmartin, Chairman, President and CEO Merck & Co., Inc. November 10, 2003.

Gulick, Luther and L. Urwick, 1937, *Papers on the Science of Administration by Luther Gulick, L.Urwick, James D. Mooney, Henri Fayol, Henry S. Dennison, L.J. Henderson, T.N. Whitehead, Elton Mayo, Mary P. Follett, John Lee, V.A. Graicunas*. New York: Institute of Public Administration.

Hatch, Mary Jo, 1997, *Organization Theory Modern, Symbolic and Postmodern Perspectives*, Oxford: OUP. pp. 63-268.

Hodgkinson, Christopher, 1978, *Towards a Philosophy of Administration*, Oxford: Basil Blackwell.

Ingram, David Bruce and Jennifer A. Parks, 2002, *The Complete Idiots Guide to Understanding Ethics,* Alpha, A Pearson Education Company.

Litterer, Joseph A. 1967, *Analysis of Organisations*, New York: John Wiley & Sons. pp. 122-131.

March, James G. and Herbert A. Simon, 1967, *Organisations, Graduate School of Industrial Administration, Carnegie Institute of Technology*: John Wiley & Sons. p. 30.

Preston, Noel, 1996, *Understanding Ethics*, Annandale, NSW: The Federation Press.
Simon, Herbert A. 1965, *Administrative Behavior A Study of Decision-Making Processes in Administrative Organisation*, 2nd ed., New York: The Free Press. pp. 20-21.
Toohey, Martin, 2004, "The Toohey Inquiry – The Restricted Report", *The Bulletin*, April 20.
Toner, James H. 1995, *True Faith and Allegiance The Burden of Military Ethics*, Lexington: The University Press of Kentucky.
Torrance, Iain, 1998, Ethics and The Military Community, The Strategic and Combat Studies Institute, *The Occasional Number* 34, July.
Tsoukas, Haridimos and Christian Knudsen (eds.), 2003, *The Oxford Handbook of Organisational Theory Meta-Theoretical Perspectives*, Oxford: OUP.

In: Contemporary Ethical Issues
Editor: Albert G. Parkis, pp. 21-38
ISBN 1-59454-536-7
© 2006 Nova Science Publishers, Inc.

Chapter 2

BIOETHICS AND NEW AGE

Victor Tambone
Dept of Anthropology and Applied Ethics
University Campus Bio-Medico,
Rome, Italy .tambone@unicampus.it

ABSTRACT

The phenomenon of New Age can be hardly outlined. In fact it is difficult to give a concise explanation of it. Through the *via negationis* (the defining way that goes on through the enunciation of what the phenomenon to be explained "is not"), can be said that it is not a movement, it is not a religion, it is not a sociological or anthropological theory and it is not, in precise meaning, a new political system. But at the same time we can identify new age movements, a new age religious view, new age sociological and anthropological theories. We think that the first defining character of the NA is its alogic-subjectivist's constitution. In this regard we completely agree with Introvigne when he points out that the main and deeper phenomenon's root is the epistemological one, even if we think it is a new kind of emotivism (subjectivist, by definition) more than a subjectivist voluntarism.

WHAT DO WE MEAN BY NEW AGE

The phenomenon of New Age can be hardly outlined. In fact it is difficult to give a concise explanation of it, it is wrong to reduce it to a specific view of man (and consequently of the *human*[1]) and it's pervaded, in an almost programmatic way, by a certain unintelligibility that make it difficult to understand its internal logic.

Through the *via negationis* (the defining way that goes on through the enunciation of what the phenomenon to be explained "is not"), can be said that it is not a movement, it is not a religion, it is not a sociological or anthropological theory and it is not, in precise meaning, a new political system. But at the same time we can identify new age movements, a new age religious view, new age sociological and anthropological theories. These features make the New Age able to permeate other human activities – because practically invisible - that are actually in contrast with it: in this way the internal "alogical" contradiction becomes a "new age" specific feature.

The first characteristic of the phenomenon that we are talking about is to have not "strong" logical connotations that imply the exclusion of the conflicting positions: thus syncretism, at truth's expense, becomes part of the *humus* in which the New Age (NA) develops.

Such a foundation rejects the concept of human knowledge that puts intellect before feeling[2], the definition of the truth as "intellect and things adaptation"[3], the definition of the human activities according the fundamental ethical category of good or evil[4], and the Popper's science and knowledge view (founded on critical rationalism).

Therefore the hub of the NA structure is the feeling, that in a religious dimension appears as a universal language that is able to involve a lot of people since *"this essential law subsists: every finite spirit believes in a God or in an idol (...). In theory, every finite good can go inside the being or values' absolute sphere that are object of an infinite tension"*[5]. Tension to the transcendent is the drive, the finite goods are the embodied disposition, but the choice directly

---

[1] In Donati's thought, the correct recovery of the human/non human category is one of the main points to reach a re-humanisation of medicine and to answer the challenge coming from the extreme technicalisation of our society. In this regard see Donati P., Sulla distinzione umano/non umano, il Mondo 3, n. 2, December 1994, pp. 158-77.

[2] In this way emotiveness appear as a gnoseological attitude in line with the new age.

[3] It's the realist gnoseological vision of aristotelian-thomist kind.

[4] *"The New Age has no objection to laying its cards on the table and to declare its absolute incompetence about the main problems of the world and that create the big utopias of the history. It is rather founded on the void of the actual moral and epistemological situation" "* (Terrin A., Il senso del bene e del male nel New Age, in Raveri M., Del bene e del male, Marsilio Venezia, 1998, p. 287).

[5] Scheler M., La fenomenologia essenziale della religione, quoted in Borghello U., Liberare l'Amore, Ares Milano, 1998, p. 57 (our translation).

derives from feeling and not from intellect, reversing, in so doing, the intrinsic order that makes those goods authentic goods for man and for the actual individual. Such a dynamics will be particularly acceptable because it will be not in contrast, by definition, with actual passions since they themselves will become the yardstick and motive of choice.

In this view, the acts, rituals and objects non-codification is an important factor, essential to the realization of the whole cultural plan we are dealing with. Just because the finite goods must be the embodiment of the religious feeling considered from subjectivist viewpoint, those goods cannot be valuable in their selves and consequently they cannot be proposed to others as goods. On the contrary, they will be the personified expression of the individual feeling that cannot be codified and that are continuously changing. The human act is understood, according to the Hume's thought, as the way the practical reason manages to satisfy the wish[6]. In this case the religious act will be the way the practical reason chooses to embody the religious feeling. From this point of view, the possible community acts are just a manifestation of the social instinct[7] or of the relation's desire intrinsic to man, also at instinctual level, but it can hardly become donation and, thus, sacrifice capability to serve the neighbour.

All this, in the actual dehumanisation dynamics where the technique is gradually substituting the relation, will lead to an apparently different phenomenon, but in fact only consequent to NA – that somebody calls "Next Age" – in which the main difference is a stronger trend towards subjectivism. We think it is not correct to say that the shift from NA to NextA is a subjectivism radicalisation because, as we have just argued, the NA's root is absolutely subjectivist. On the other hand, we think that in the NextA the consequences of such a subjectivism are committed and the subjectivism is made more evident by the crumbling of some pseudo-communitarian practices that mimed a certain social or solidly trend. In other words, if the NA supports a new era of happiness for all, the NextA goes behind the evident historical failure of this theory assuring the possibility to go inside a personal and individual New Era of happiness through a lot of possible techniques[8], applied to their selves and for their selves.

---

[6] Caffarra C., Introduccion General, in Aquilino Polaino-Lorente, Manual de Bioetica General, Rialp Madrid, 1994, p. 26.

[7] With "social instinct" we mean the natural tendency of man to live in society.

[8] It will be interesting to deepen the fact that the NextA can be interpreted almost like an extreme technologisation of the religious feeling that, for that reason, will be absolutely dehumanising and apersonal.

Here the NA can be seen as a sort of a "spiritual life Super Market". Igor Sibaldi[9], following this line, thinks that, the time of the Churches being passed by now, the Market is now the main spirituality guide. This is considered as subjectivism expression in this field; an individualism that must be taken into consideration if one wants to understand the reason of the decline or the success of a specific spirituality. In this way the spirituality is understandable only through the market rules; it becomes a matter of the market in which to offer products on the basis of the possible customers wishes taking into consideration, Sinibaldi points out, that "it is anyway a very changeable material. Angels last one year and a half, then something else is needed". Super Market as possible choice of the sets-up to believe in, of the supernatural meeting to be experienced, of the mystic sensations to live. Super Market that makes the heresy a reason for existence and thus that shapes *a heresy redemption period: (in fact) in Greek heretic means the one who makes a choice, the customer.*

As explained up to this point, we think that the first defining character of the NA is its alogic-subjectivist's constitution. In this regard we completely agree with Introvigne when he points out that the main and deeper phenomenon's root is the epistemological one, even if we think it is a new kind of emotivism (subjectivist, by definition) more than a subjectivist voluntarism.

FROM THE HISTORICAL POINT OF VIEW THE PHENOMENON FOLLOWED THIS CHRONOLOGY[10]:

- **XVIII Century:** during this century important experiences are realized that are a remote preparation of NA. They are the occultist and mystery practices that on one hand develop within the spiritism and, on the other hand, allow the eastern mysticism and Tibetan mythologies, through theosophical society, to go inside the western culture. Then it goes so far as the Rudolf Steiner's Anthroposophy and traditions, as the human kind's "Annales Akashiques", that will be in the NA.
  Such esoteric culture has the following features[11]

---

[9] Sibaldi I., I maestri invisibili, Mondadori Milano, 1998.
[10] We have written this chronology referring especially to the studies on this subject by Jean Vernette, Massimo Introvigne, Teresa Osorio Gonsalves and Gianfranco Ravasi.
[11] See Faivre A., Access to Western Esoterism, Sunny Press, Albany 1994, pp. 10-15.

a) The visible and invisible universes are linked through influence at all levels (for examples also the human body and the planets);
b) The nature is a living being animated by a force that man tries to control;
c) Through imagination and mediators, man can enter into contact with higher worlds;
d) A route to reach the knowledge of a redeeming revelation (Gnosis) is proposed;
e) A secret wisdom tradition is looked for to be the interpretation key of any esoteric act;
f) There is an initiation from master to follower.

- **1875:** with the theosophical society's foundation there is an important attempt to link the only just outlined tradition with modern culture developed around scientific discoveries. In this way they try to reject Christianity to seek a universal religion, that includes all the faiths, and the man's hidden potentialities. Surprisingly, some experiences deriving from the Theosophical Society try to reconcile esotericism and Christianity[12].
- **End of XIX Century:** The concept of New Age as "Aquarium Age" develops. It follows the Pisces Age, characterized by the Christianity beginning and development. This idea originates in the esoteric-occultist ambient and it refers to authors like Paul Le Cour (1871-1954) and Alice Ann Bailey. The New Age should imply some deep changing: the end of Christianity, another coming of Christ and an age of full earthly happiness. In 1937 Paul Le Cour publishes his book *"L'Ere du Verseau"* where, on the basis of the astrological theory that maintains the sun's sign change every 2169 years, he indicates that the Pisces Age (begun on 21st of march of the Christian age) is going to finish and it is about to starting the aquarium age that will be characterized by the abundance.
- **1960:** starting from 60s, the Theosophical Society inspires 2 centres, one in Scotland (Utopian Community of Findhorn) and the other in California (Institute for the development of human potential of Esalen). These places become the points of reference and development of the NA with the aim to create *"a new man, setting free his body, increasing his*

---

[12] For example the C.W. Leadbeater's liberal Catholic Church and Alice Bailey's Arcane School or the Steiner's Anthroposophy.

knowledge"[13]. In **1968** people come into contact with this phenomenon through the famous musical "Hair", where the ambient of youth counterculture merges with the aquarium mythology.

- **1980:** Marylin Ferguson publishes the book *"The Aquarium Conspiracy"* where the political, social, scientific, psychological, educational and religious NA's spins-off are even more outlined, trying to collect all these tendencies in a common conscience. As a matter of fact, it would constitute a conspiracy to facilitate the coming of the new age. From that, the idea of NA as a new paradigm, as a new vision of the world originates.

- **1980-90:** the NA expansion on a large scale. During this ten-years period there is a wide diffusion or application of the NA paradigm leaded especially by spiritualist and theosophical groups. "Networks" around the world are constituted and *a sense of community between prophets and little groups was being created* (developing) *the image of a growing movement able to permeate the society going over the very follower circles.*

- **The present crisis:** by now many authors, J. Gordon Melton and M. Introvigne between them, consider that NA is passing through a crisis and that it is letting NextA pass. However it continues to be an active phenomenon in Europe and, according to many, it is very viable. At the present state of things I think that, even if as cultural movement it can be motionless or fossilized, its consequences on the mentality and on the practice are very active: this is the reason why it seems interesting to me to deepen its relations with the present problems of Bioethics.

These are the main steps that lead to the present "NA Paradigm". The several factors that can be found in the forgoing chronology are needed to understand what we mean when we talk about NA as a "Net", a net of different elements but linked by a new way to consider and interpret life and man. It's just in the logic of the applied paradigm that the influence of such "sweet aquarian conspiracy" on our society can be studied and better understood.

---

[13] Vernette J., La Nuova Era, Pontificia Università S. Tommaso d'Aquino Roma, 1996, p. 10 (our translation).

WAY OF NEW AGE EMBODIMENT

Talking about real "products" that are realized in such a Net, could be re-lead us to a theoretic analysis of the phenomenon and this is obviously to be avoided. Thus, we will offer in the continuation only some tangible examples, useful to facilitate the comprehension of what it is stated up to now.

Literature, through authors from all part of the world, publishing houses able to realize great advertising campaigns and especially through a strong and wise synergy (typical of a Net) between different mass-media instruments, is successfully transmitting the NA paradigm. The book turns into the movie, the movie turns into the gadgets, toys, role games and much more that gradually transmit a particular idea about life, death and universe.

Such a tendency can be observed at various levels.

Some time ago I was talking with a friend of mine whose father died a short time before. All the family lived serenely the loss, also the 4 years old daughter who unconsciously set her grandfather's death in a vitalistic NA-style view. In fact when the father asked her if she understood what happened to her grandfather, the child answered: "of course, it happened the same to the Lion King!" Simple, reassuring, in line with a life conceived as a cycle with shapes that have not personal subsistence but that is continuously reincarnated.

The work by Robert Mawson titled "Lazarus child", is the last best-seller that the *Publisher Weekly* defines "a New Age thriller shrewd but topical". It is the story of a little girl who goes into an irreversible coma and of a very alternative medical doctor who tries to bring her back to conscious life with means, may be questionable, that make a small opening in the mysterious, pseudoscientific and hidden world of the *"out of body"*, life beyond life experiences. This explains, in a certain mystery way, analogies and connections the NA is always looking for, and that always offers as knowledge and true humanity's discovery means. The italian analogous of the Mawson's book, is *Una porta di luce* written, almost at the same time, by Mario Biondi and published by Longanesi. Obviously it is not a new kind but it is a product in great demand and, thus, a best-seller. You have only to say that the Mawson's work made him 1.5 millions dollars only by US copyrights and it has been translated simultaneously into French, Italian, Spanish and Dutch.

Works by Anthony De Mello, both originals and texts not authorized by the author, are another example of literature with some aspects of NA philosophy. They deal with the relationship between man and God and the creation's religious vision. The theme of an impersonal God is the most steeped by NA influence, even if probably in an unconscious way. This God is practically indistinguishable from creation and thus he cannot be really known. He can only be met in an intuition that end by clashing with void, with a cosmic omnipresent reality: *"God has nothing to do with the idea you have about him... There is just a way to know him: not to know him"*[14]. According to De Mello, nobody can help us to meet God, like nobody is able to find a fish in the ocean[15]. Furthermore he states that we are not the same thing with God neither we are two, like the sun and its light, ocean and its waves are not the same thing but neither two[16]. We are arriving to a second agreement with the NA paradigm: the wish to reject any religion to come to a syncretistic and universal religious view. The fundamental concept of such religiosity is the flash of inspiration essential to know the good and the evil. The one who has a faith in which he... believes, is a fanatic who makes it difficult to approach the redeeming truth. The same author says: *"if you live as a communist or a capitalist, as a Moslem or a Jewish, you live your life in a preconceived and tendentious way: this is a barrier, a layer of fat between the Reality and your spirit that is not able to see and touch it directly any more"*[17].

Cinema is a privileged place to study the NA paradigm. Through this wonder vector of ideas, the typical subjects of the "Aquarian conspiracy" are approached; the existence of a sort of a parallel world beyond the death; the existence of Extraterrestrials as Brothers we were separated from for a long time but who will be able to show us the way to come back to the original happiness; Angels, not in the Christian version (remember the masterpiece by Frank Capra "It's a wonderful life") where they help men to live their life to the good, but in a new version (refer to "Michael") aimed at reaching welfare, the final event, the apocalyptic "Armageddon", etc.

A careful and deep reading of the NA cinematography leads us to focus on an important and clarifying point[18]. Man continues to experience a great anxiety about his salvation, about the knowledge of what transcends him.

Classically, and for many centuries, human intellect referred to the four transcendental categories of the human being, i.e. the One, the Truth, the Good, the Beauty.

---

[14] De Mello A., Istruzioni di volo per aquile e polli, Piemme Casale Monferrato, 11 (our translation).

[15] See De Mello A., Un minuto di saggezza, Piemme Casale Monferrato, 77; Messaggio per un'aquila che si crede un pollo, Piemme Casale Monferrato, 115.

[16] De Mello A., Un minuto di saggezza, Piemme Casale Monferrato, 44.

[17] De Mello A., Chiamati all'amore, Piemme Casale Monferrato, 62 (our translation).

[18] Here I refer to concepts expressed by Professor J.M. Galvan from the Università Pontificia della Santa Croce, at the presentation of the book "Il dio della California" by C. Siniscalchi, Ente dello Spettacolo Editore Roma, 1998, which we suggest to deepen the subject of New Age and Cinematography.

As it is known, a wide part of the contemporary culture rejected Truth more and more radically, theorizing an agnostic Gnoseology that came true in a vital attitude of gnoseological relativism. However, when the way to the Truth is destroyed, rapidly the concept of Good as moral truth is destroyed as well. Thus together with the gnoseological relativism rises the ethical one. This phenomenon, once removed the ability to know the truth, removes the ability to know the good. Consequently the man's native tendency to the transcendent pursuit (and to the comprehension of the self) must be concentrated on the last transcendental: "the Beauty", got away from the Truth and the Good, that will be found the last hope of salvation on the Aestheticism. In other words, we will not contemplate any more the Truth but the beauty, we will not pursue the Christ's contemplation (I am the Truth) but the crystal's one. This is the clue to understand many of the manifestations of this millenarianist paradigm.

According to some, a first example of the connection between literature and cinematography explainable as a part of the "Net" is the Star Wars saga[19]. Some authors think it is possible to see in the Chinese culture the real root of the Lucas' philosophy (see, for example, F. Sisci, Quella forza nasce in Cina, Specchio n. 175, 29.5.99, pp. 77-78).

Here the concept of "Power" would be the translation of chinese "Gong" (also said Kung), hub of the martial art for antonomasia, exactly Kung-fu or Gong-fu; the powers of the Jedi knights would be the same that the martial arts masters of the chinese tales have got; the fighting techniques, with the "laser sword" status symbol, follow the art of Kendo; Luke Skywalker's training also includes the qi gong (the power of the air), a chinese concentration technique useful to increase the strokes' efficacy. Furthermore it is to be noted, on the other hand, that the Saga is a vector of "strong" principles; that it clearly defines the good and the evil; that it lead towards a coherent moral interpretation of the events connection in a coherent story not alogic at all; that the idea of self-perfectioning goes also through effort and sacrifice and not only through the feeling following; that Dart Maul is intentionally "depicted" like a devil in order to be linked with the traditional concept of evil's source or evil's agent. In conclusion we think it is not correct to identify the Star Wars Saga as a "New Age" product.

---

[19] The Empire, evil at its pure status, clashes with the residual resistance of a Rebel group guided, sometimes in a conscious way, sometimes in an evident way, by a "Power", beneficial vital principle, cosmic and universal. The key to the first part of the story is the "re-awakening", the consciousness taking by the last Yedi knight of his noble nature and of how he can become channel of the Power to change the dramatic course of events. In this great saga several elements coexist: cosmic vital principle, mystery initiation, channelling ability, reincarnation, paranormal mental powers, extraterrestrials and much more. Also the happy ending will be part of the vision guided in a way by the stars, the coming of an Age of Universal Peace.

Other examples of NA cinematography can be:

a) "Close encounters of the third kind", by Spielberg, where the extraterrestrials based religion is clearly created. Here traditional Christian concepts like Grace and Salvation, transformed in a new "extraterrestrial" view, are reproposed: the close encounter with these non terrestrial beings is a "redeeming" contact that can be realized, sometimes, leaving the body at the exact astral time and finally letting the spirit reach the other world, the happy world. In this way some tragic collective suicides recently happened can be explained;

b) Some Walt Disney's movies like the Lion King and Pochaontas. In the first one the cyclical vitalistic conception is explicitly explained by a voice out of the scene that, as a prologue and key of the story, expounds that Life is a Cycle with different forms in plants, animals... This idea is dramatically confirmed in the mysterious face-to-face encounter between the little Lion and his father, the dead King. The little lion's question is immediate and simple, it originates from this surprising contact with the non-life world: "where are you?". The answer is clear, encouraging and loaded with the power of the spirits: "I am in you!".

In this statement could be found what somebody think to be the reason of the present New Age crisis, i.e. the mean that the "paradigm" gives to the authority of the encounters based on channelling.

With "channelling" the possibility to direct information and suggestions from a guide spirit is meant: an immaterial being of a superior evolutive level. Receiving directives from this conscience living in another dimension, it would be obtained a sort of prophetic ability, an authoritative knowledge of the facts: as a matter of fact, this erodes the subjectivist relativism that sets up and supports the NA system.

In Pochaontas, instead, there is, in a way, the view of the Earth as a living being in a close and vital connection with man: the willow-grandmother, we think, is an efficacious exemplification of it.

At the musical level, the starting point on a large scale has certainly been the famous musical "Hair" in which the Aquarium Age was announced, in a way, through the well-known song "Aquarius". However the characteristic of the NA music is not to talk about itself but to carry a manner of conceiving music based, as we have already said, on the "beauty" paradigm (not linked with "truth" and "good") and on its ability to evoke, through many different styles, mystery or, anyhow, indefinitely spiritual contents. As a matter of fact, *"the book of the new* *age harmony gives the composer a practically infinite freedom. The only condition is about the theme of his composition that must be, in a way, the beauty"* [20].

The roots of the NA music can be found in the Muzak[21], progressive Folk[22], Space Jazz[23], electronic Rock and in many other kinds that come together in the World Music. It is inspired by the Folk from all part of the world and it is the main reference point (according to Scaruffi) of the NA music. In this context interests in joining western and Indian music (Pink Floyd) with African (David Fanshawe) or Native American music (Carlos Nakai, Jessita Reyes and Dik Darnell) are developed.

The use of instruments in the NA music has a particular meaning that can easily be found in the use of harp, percussion and ethnic instruments. The harp is particularly suitable for transmitting messages of emotional spirituality, sharing with nature and abstractionism. This instrument is considered particularly important, *"NA musicians found in the harp the ideal vehicle to put the wild nature in the domestic dimension. And they have simply tried to attract a certain magic and timeless that the soul strings spreads in melodies. The new age musician feels himself to be programmed to passively follow the sound from harp rather than trying to tame it. The harp's sound is one of that which echo in the human soul deepness, one of that which more than others can lead man to understand who is he"*[24]. Percussions are frequently used with seeming hypnotic purposes. Ethnic instruments are exploitable to excite unusual sensations. Voice as well is used to transmit emotions in the wake of the Tibetan and Sufi singing.

---

[20] Piero Scaruffi, Enciclopedia della Musica New Age, p. 22 (our translation) available on http://www.scaruffi.com/avant/na1.htlm.

[21] Muzak is the soundtrack music, a music not to be listened but to follow images. In Chip Davis we can found an interpretative reference. It is an extremely figurative kind of music that describes moods, emotions, places, etc. George Melachrino, Mantovani, Percy Faith are inspiration sources in this genre. In 1966 "One Stormy Night" by the Brad Miller's Mystic Moods Orchestra is produced. It is a particularly useful product to understand the muzak role in the NA paradigm. "One Stormy Night" wants to represent the union with Hippy sensibility. On the same line is the soundtrack of the TV series "Twin Peaks", called "elevator noir".

[22] Right in the Folk genre, through the unappreciated activity of Sandy Bull and others, there is the first time fusion of eastern, western and African rhythms. A further step is recognizable in the "American Primitive Guitar" movement by John Fahey that diverges from Nashville rules. Simultaneously in Great Britain Celtic music is rediscovered even if it looks for new kinds of expression and of fusion with other music traditions. Syncretism seems to be a typical NA character even in the musical dimension.

[23] It is a genre coming from the evolution of Fusion that tends to depersonalise jazz, trying, like it happens in muzak, to transmit an experience, an emotion, an atmosphere. We recall Terye Rypdal and Jan Garbarek.

[24] Scaruffi P., Enciclopedia della Musica New Age, entry "harp" (our translation).

In this new musical paradigm an instrument in particular is removed: it is the electric guitar even because it is associated with rock music, which has too much exalted a discordant, conflicting, and aesthetically decadent kind of life.

Beyond the specific, technical considerations, it seems important to me to underline that NA music is not only a particular music, played in a particular way and with particular contents, but it is also a syncretistic way of music listening that gives an emotional reinterpretation of many musical genres strongly characterized by content and message. Let's give an example. Virgin commercialised a CD titled "Mystica" with 18 tracks joined by being compositions of "music for the soul". It could be surprising to find together Gregorian with ceremonial American Indian music, MacKennit, who set San Giovanni della Croce to music, with a classical example of World Music like Adiemus. But instead of surprising us, this record product confirm us to be a clear example of what the NA paradigm wants to realize in connection with the spiritual phenomenon's relativization and emptying (with regard to contents).

Other ways of NA realization can be: a) the use of astrology by Companies to plan their activities and select the staff (company's astromethodology); b) astrological financial advice (six months financial forecasts in Italy cost approximately 150 euro); c) Yoga and transcendental Meditation courses for company's staff in order to optimise its work.[25].

BIOETHICS AND NEW AGE

It could be said, according to what has been written up to this point, that the NA influence on bioethics is just a mentality matter. This is true only if we interpret the "mentality" like a way to understand life and man, based on anthropological and philosophical well-defined concepts. For that reason we think important to identify anthropological NA concepts that can influence practical bioethical decisions.

In fact Bioethics is not a simple pragmatic decision-making formulation uprooted from fundamental philosophical reflection but, on the contrary, it is a rigorous application of it even if not always definite. Thus, in the following we will analyse the NA point of view with regard to 3 subjects fundamental for the definition of present tendencies, ethically very important, in biomedical practice:

---

[25] An experiment in this sense is the week in the New Mexico desert offered by some companies to their executives, or the *Managing self* course organized by IBM for future European managers.

a) the idea of life b) the idea of death c) philosophical formulation of medical science.

THE IDEA OF LIFE

Due to the very relativist character of the NA paradigm, it is difficult to identify its idea of life. On the other hand there are constant elements like, for instance, the man and nature community, the cyclical aspect of life and the existence of a unifying cosmic power. According to these characters there is a human life view as a cyclical Vitalism.

Such an attitude so largely diverges from the personalist view of Man as to reject, as a matter of fact, a personal difference in men who would be, in unity with nature, different kinds of manifestation of the same vital, universal principle.

This kind of view implies an important consequence about the concept of Contragestion[26] coined by Baulieau. This new word, according to the author, is necessary because the term "abortion" is a violent term and it has a negative connotation. The aim of the new word, instead, is to avoid systematic mental block that are provoked by a charged with emotional terminology. The ideological plan behind the term "contragestion" is very clear: the cultural suppression of the abortion consciousness. The question is not only to remove the sense of guilt associated with abortion, but also to reduce the human life transmission to something essentially physiologic and indeterminate. In fact, in this context, fertilization in not a particularly important event of a new human being's birth any more. It is just a further event, preceded by many other events and followed by the indenization process, devoid of its specific meaning. Starting from this point, Baulieu passes to define the life's generation as a continuous process, stating that *"However fertilization is not the unique determining event in the conception of a new human being. (...) Thus the generation of life, and human life specifically, is a continuous process that involves interdependent sequential events and that cannot be attributed uniquely to fertilization"*[27]. In other words, contragestion's phenomenon, or the early anti-indenization abortion, requires rejection of the syngamy's specific meaning as a new human life-founding event; on the contrary it requires a vitalist theory as a sufficient anthropological foundation.

---

[26] Contragestion is an abbreviation of contra-conception. Referring to the most widespread meaning of the word "conception", the term contraception is synonymous of "prevention of fertilization".

[27] E.E. Baulieu, RU 486 as an antiprogesterone Steroid, in Jama, (October 6th 1989) 1813.

This observation is strengthened by the fact that also R.G. Edwards, who realize IVF-ET, made a similar anthropological consideration (foundation) going as far as stating (nearly with the same words used by Baulieu) that *"fertilization does not begin life. Life is a continuous (...)"*[28]. Even in this case, the justification of an act that involves a high number of embryo's sacrifice requires supporting a vitalist view that denies the syngamy's value.

In this way the NA human life paradigm gets to the heart of matters like abortion and IVF.

The Idea of Death

We have already hinted at the fact that NA refers to a parallel world, the one beyond the death, which is connected in various ways with man's life. What does not exist is undoubtedly an eternal life where we will receive the punishment or the reward for how we acted in our earthly life. Reincarnation and transit, through the death, to a perfection status, are ideas present in the Aquarium anthropology and eschatology.

In fact, the reincarnation concept is coherent with the vitalist (apersonalist) view we have only just talked about because it is against the idea of the human being as a unique and personal individual and it comes from believing in the existence of an only universal vital Power.

Obviously reincarnation, or death as extreme channelling towards a perfection state, deals decisively with reflections about euthanasia and assisted suicide. If physical life is without value as regard the person, even because the very concept of personhood is rejected, the choice to kill himself does not have, *per se*, any ethical relevance, but it is turned into a mere convenience question. On the other hand, such an interpretation of death leads to the logical consequence, in many cases at least, of considering euthanasia as the solution of an absurd and not worthy life from the "beauty" paradigm view.

Religion UFO and Cloning

As we know, one of the NA features is the so-called "channelling", that is the presence and the connection, in a parallel world, with guide spirits. Extraterrestrials (the Elohim) embody this idea so that the role of the rational

---

[28] R.G. Edwards, In Vitro fertilization and Embryo Transfer, in Annals of the New York Academy of Sciences, (1985) Vol. 442 565.

consciousness can be transferred to their wise revelation that is source of practical-normative flash of illumination[29].

An example of the dismantling capacity of such a thought is the miserable experience that Raelians[30] are offering about human Cloning. Even if such a phenomenon may seem to be ridiculous, we must be able to twig that it is a manner of preparing a specific part of the public opinion to conceive cloning as a positive event, through an action of emotive, alogic and gnostic mould, typical NA features.

THE PHILOSOPHICAL FORMULATION OF MEDICAL SCIENCE

It would be incorrect to include in the NA classification all the medical practices known as "alternative medicine". In reality we do not include in the NA classification medical practices like homeopathy and acupuncture. These are different ways of a paradigm application that fails to refer to biology and scientific medicine in order to satisfy, as Lucien Israël states, *"customers looking for exoticism, alternativity, a new age with its ambiguities and its rituals and, furthermore, who obstinately want that not only their needs but also the inextinguishable desires are satisfied..."*[31].

When we talk about NA medicine, we refer to practices at the edge of exotericism, that search the activity of the so-called "thin energies" that would be present in nature and in man. It is the energy of crystals, colours and geometric shapes.

Considering the impossibility to clearly classify this phenomenon, nevertheless some heterogeneous kinds of therapy may be considered belonging to this context, for example the pyramid therapy, crystals therapy, medical astrology or the Touch Therapy. In some cases they are not new phenomena but they are rediscovered and re-proposed by the new NA paradigm.

The aim of the diagnostic phase is firstly to identify the energetic state of the patient through instruments like pendulum o other esoteric methods.

---

[29] Here we refer to the cults of UFO as "contactists" followers and not to the category of ufologists who don't have any religious interest in that subject. By now, in Italy, according to the CENSUR report of 4th December 2000, five are the organized groups; the Raelian Religion, Unarius (Universal Articulate Interdimensional Understanding of Science), Giordano Bruno Association, NovaConvivia and the C.O.O.P.C.O.S.M.O.

[30] Raelian Religion is the most widespread ufological cult in the world. It states that men have been created in laboratory by extraterrestrials, the Elohim. The Elohim revealed to Rael that God, soul and eternal life do not exist and that praiseworthy ones will be recreated in their planet.

[31] Israël L., La vie jusqu'au bout, Plon Paris, 1993, p. 25 (our translation).

Generally there is a tendency to underline the therapeutic ability of the "Energy": the pyramid form would tend to create particular beneficial waves, crystals would be able to emanate beneficial energy to the body, astrology would involve thin therapeutic energies. This energy would have a general body rebalancing effect.

In this field an interesting example is the Flower therapy. This technique originates from the E. Bach's thought. 38 kinds of flowers are involved, they are distilled in pure water through a methodology that utilizes the energy of the sun. According to its founding theory the flower's vibrational nature is transmitted to the water that in this way will become active to the patient's psyche that is, according to Bach, the source of the disease. A last derivation of the vibrational medicine is Bud therapy that utilizes elixir of bud or minerals treated with pure alcohol: the solution gets the chromatic energy of the stone.

All these techniques, according to some, have a "paranormal" character in common, to the point that their efficacy is lost when they are subjected to comparable tests.

It's evident the methodological point-break with scientific medicine: hypothesis does not originate in a logical procedure based on the cause and effect relation, easy to check through experimental method. On the contrary, it is an alogic, not testable and paranormal methodology.

In other words this kind of medicine is a regression to the Magic Medicine by Asclepius: that was got over by the Greek metaphysical thought based on the cause and effect connection and that is the foundation of Modern Scientific Medicine *"the scientific mentality, created by physis philosophy, made possible the constitution of medicine as science (...): medical science (and the other sciences) could originate, self-define and develop just within the philosophical mentality, that is within the etiologic rationalism that it created"*[32]. From this point of view, Evidence Based Medicine[33] appears as a particularly important way to be followed in order to avoid on one hand the extreme technicality of medicine and, on the other hand, its regression to a magical conception rejecting all the scientific progresses happened in two millenniums.

---

[32] Reale G., Antiseri D., Il Pensiero Occidentale, Editrice La Scuola Brescia, 1985, Vol. I p. 82 (our translation).

[33] Evidence Based Medicine is a clinical methodology where "physician decisions, assisting individual patient, must be the result of the combination between experience and scrupulous, explicit and judicious use of the best available scientific evidence" (Sackett DL, Rosemberg WMC, Gray JAM, Haynes RB, Richardson WS, Evidence Based Medicine: what is and what isn't, BMJ 1996; 312:71-72 quoted in the italian translation by Cartabellotta A., et al., Evidenze scientifiche e Tests Diagnostici, Medic 1997; 5:117) (our translation). We remind that term "Evidence" is here used with its english etymology that means "proof".

The connection between the specific clinical case, the natural history of the disease and the local Medicine; the reference to the clinical data objectively detectable through objective and instrumental observation; the comparison with the scientific knowledge coming from clinical experimentation; utilization of metanalysis and clinical trials make the diagnostic and therapeutic process a rational and comparable process which tends to a greater and greater approximation to the disease in question. Criticism is always possible, but as a coherent rational position starting from new data that, in a way, will lead to a greater approximation to the scientific truth that always remains the intentional reference point.

Just in this way it is possible to structure a rational diagnostic and therapeutic action, with a specific scientific methodology that originates in the testable facts.

Due to what we said about the metaphysical origins of scientific medicine and the danger of its extreme technicalization, this important role of EBM requires that knowledge of classical metaphysical thought and Donati's relational model be increased.

The knowledge theory which EBM makes use of, originates in a critical rationalism that, from the gnoseological point of view, admits the possibility to know just at the experimental level.[34]. This Gnoseology is not able to justify itself and denies other knowledge means at ethical, relational and contact with reality level. EBM will loose part of its mean as "ridge" for the NA alogic emotivism as long as this problem will not be solved. This is because it will tend to reject a part of man, the same part that man cannot reject in his existential experience. The matter is to radicalise trust in the human reason without to reduce it to the function of experience's technical-scientific interpretation.

The bioethical effects of this methodological equilibrium are very important. In fact, if the contact with the Scientific Medicine is lost, the assiological reference, that the Belmont Report synthesises in the well-known principle: "Bad Science = Bad Ethics", will be lost; guidance of Good Clinical Practice wont be valid any more; it will not be possible to refer to professional virtues based Ethics and the virtue of Prudence in clinical practice will lose any possible normative reference. In other words the question will be the ethical confusion resulting from the truthful and rational reference loss.

---

[34] See, Possenti V., Razionalismo Critico e Metafisica, Vita e Pensiero Milano, 1996.

For this reason it seems desirable to us the widening of this subject from a transdisciplinary point of view, able to find solutions to reach an entirely Rational Clinical Methodology, able to draw lesson even from Classic Realism.

In: Contemporary Ethical Issues
Editor: Albert G. Parkis, pp. 39-61

ISBN 1-59454-536-7
© 2006 Nova Science Publishers, Inc.

Chapter 3

CONSENT AND SENSIBILITY

Alasdair Maclean
The University of Dundee,
Scotland, UK

ABSTRACT

Consent is too complex to be understood fully in any one theoretical model'.[1] In this article my approach is that a core concept may be isolated and it is the peripheral attributes that are the seat of most disagreement. I intend to take a fresh look at consent and to cut away the excess flesh, which should make it easier to distinguish those things directly required by consent from those that need independent justification. In the first part I will consider the concept of consent. In the second part I will examine the moral basis that justifies consent and in part three I will discuss the implications of these prior discussions for the attribution of legal responsibility. The topic is voluminous and so my discussion must necessarily be selective. Its primary aims will be to reveal hidden or underlying assumptions and to explore the relationship between consent, autonomy and the responsibility for outcome

There is nothing new about the basic idea of consent. It was certainly recognised by mediaeval courts as a defence to a charge of assault[2] and seems to

---

[1] Alderson, P. Goodey, C. "Theories of consent" (1998) 317 *British Medical Journal* 1313 at 1315.
[2] *Bridelyngton v Middlilton* (1388) CP 40/512, m.124, abstracted in: Baker, J.H. Milsom, S.F.C. *Sources of English Legal History: private Law to 1750* (1986) London: Butterworths at 322. This case concerned two youths engaged in rough horseplay and the term used by the court was

have been 'an important issue' for both Greek and Byzantine authors.[3] Alexander the Great's physicians, for example, were apparently reluctant to treat him without explicit permission.[4] In *Slater v Baker and Stapleton*, the earliest English legal case to find a surgeon liable for failing to gain the patient's consent, the plaintiff's broken leg had been poorly set and healed crookedly. The defendants re-broke the leg and used a steel device to try and straighten it. In this time before the invention of anaesthesia, the court held that:

> it appears from the evidence of the surgeons that it was improper to disunite the callous without consent; this is the usage and law of surgeons: then it was ignorance and unskilfulness in that very particular, to do contrary to the rule of the profession, what no surgeon ought to have done; and indeed it is reasonable that a patient should be told what is about to be done to him, that he may take courage and put himself in such a situation as to enable him to undergo the operation.[5]

These early examples highlight three possible roles for consent to medical interventions: as a defence against complaint; to allow the patient to know what to expect and prepare himself for the ordeal; and to facilitate the patient's cooperation. As recently as 1993 a leading English judge confirmed two of those functions, stating that the legal purpose of consent was to protect the doctor from unwarranted claims and the clinical purpose was to gain the patient's cooperation.[6] Over the last 60 years (since the Second World War and the Nuremberg trials), however, an additional role has been brought to the fore: that of protecting the patient's right to self-determination or autonomy.[7] It is this change in focus that perhaps goes some way to explain why some commentators claim that, in the healthcare context, consent is a new phenomenon.[8] Further

---

'common assent' rather than 'consent'. Nevertheless it illustrates that the courts accepted the basic idea of consent as a defence to a battery charge.

[3] Dalla-Vorgia, p. Lascaratos, J. Skiadas, P. Garanis-Papadatos, T. "Is consent in medicine a concept only of modern times" (2001) 27 *Journal of Medical Ethics* 59.

[4] Dalla-Vorgia, p. Lascaratos, J. Skiadas, P. Garanis-Papadatos, T. "Is consent in medicine a concept only of modern times" (2001) 27 *Journal of Medical Ethics* 59 at 60.

[5] *Slater v Baker and Stapleton* (1767) 95 ER 860 at 862.

[6] *Re W (A Minor) (Consent to Medical Treatment)* [1993] Fam 64 at 76.

[7] I will explain the distinction later.

[8] See e.g. Polani, P.E. "The Development of the Concepts and Practice of patient Consent" in: Dunstan G.R. Seller, M.J. (eds) Consent in Medicine: Convergence and Divergence in Tradition (1983) London: King Edward's Hospital Fund 57-84; Habiba, M.A. "Examining consent within the patient-doctor relationship' (2000) 26 Journal of Medical Ethics 183.

explanation comes from the concurrent increase in legal claims, primarily involving a failure to disclose information relevant to the consent decision. The situation is complicated since many of these legal cases, reflecting the development of the law, are claims for compensation when an undisclosed risk materialises and causes physical injury.[9] Thus, consent is used in these negligence-based claims as a tool to allocate responsibility for outcome rather than a protection of patient autonomy. The water is muddied further by the development of arguments that focus on the positive right of autonomy, resulting in a new role for consent: to protect patient choice.[10]

In their article that describes five distinct models of consent, Alderson and Goodey conclude that: 'Consent is a strong concept in being so versatile and durable, but it is vulnerable to conflicting interpretations and rejection as a worthless ideal... Consent is too complex to be understood fully in any one theoretical model'.[11] In this article my approach is that a core concept may be isolated and it is the peripheral attributes that are the seat of most disagreement. I intend to take a fresh look at consent and to cut away the excess flesh, which should make it easier to distinguish those things directly required by consent from those that need independent justification. In the first part I will consider the concept of consent. In the second part I will examine the moral basis that justifies consent and in part three I will discuss the implications of these prior discussions for the attribution of legal responsibility. The topic is voluminous and so my discussion must necessarily be selective. Its primary aims will be to reveal hidden or underlying assumptions and to explore the relationship between consent, autonomy and the responsibility for outcome.

THE CONCEPT OF CONSENT

It might be thought that the basic idea of consent is relatively straightforward and certainly consent is often written about as if the basic concept is uncontested.[12] However, just as the more peripheral attributes of the concept are

---

[9] Dworkin, R.B. 'Getting what we should from doctors: rethinking patient autonomy and the doctor patient relationship' (2003) 13 *Health Matrix* 235 at 240.
[10] Alderson, P. Goodey, C. "Theories of consent" (1998) 317 *British Medical Journal* 1313.
[11] Alderson, P. Goodey, C. "Theories of consent" (1998) 317 *British Medical Journal* 1313 at 1315.
[12] See, for example, Dickens, B.M. 'Dimensions of informed consent to treatment' (2004) 85 *International Journal of Gynecology and Obstetrics* 309-14. This is an observation rather than a criticism.

debated so are the attributes at the very core of consent. It is here that my exploration begins.

The central notion of consent has three features: what it is, how it works, and what it does. Regarding the first feature, the main possibilities are that consent may be seen as a state of mind, as an act of communication or as some combination of both. Onora O'Neill has argued that consent is a state of mind. Thus, she suggests that: 'A reasonable starting point is to note that consent is a *propositional attitude*, given in the first instance not to another's action, but to a proposition describing the action to be performed'.[13] By starting with this assumption she is able to argue that, like other propositional attitudes, consent is opaque and thus, because communication is incomplete and imperfect, it is impractical or perhaps impossible to achieve.[14] Instead of trying to realise this metaphysical consent,[15] O'Neill argues that it is better to focus on a more 'genuine consent', which requires that patients 'are neither coerced nor deceived, and can judge that they are not coerced or deceived'.[16]

O'Neill is not alone in her view that the essence of consent is a state of mind. Hurd, for example, suggests that 'consent must essentially constitute an act of will – a subjective... [and] purposive mental state possessed of propositional content'.[17] Likewise, Alexander states that: 'consent... must be the exercise of the will and, thus, a subjective mental state'.[18] The problem with viewing consent solely as a state of mind is that, if consent does not need to be communicated then how does the actor know when it is permissible to act? If the actor's behaviour is not to be condemned then the act must be justifiable. If consent is crucial to that justification then, although the person acted upon may have formed the appropriate mental state so that the act itself is permitted, it is arguable that the actor's behaviour may still be condemned. In the absence of some indication that the other is consenting, whether express or implied, the actor has no reason to believe that the act is permitted. Unless the actor is a mind reader he cannot know whether the other person has formed the necessary attitude to his proposal. To carry out the act regardless, where a simple enquiry would have resulted in the granting or withholding of permission, is to act recklessly in a way that focuses too closely on the act at the expense of the relationship between the two parties.

---

[13] O'Neill, O. 'Some limits of informed consent' (2003) 29 *Journal of Medical Ethics* 4 at 5.
[14] O'Neill, O. 'Some limits of informed consent' (2003) 29 *Journal of Medical Ethics* 4 at 6.
[15] My term.
[16] O'Neill, O. 'Some limits of informed consent' (2003) 29 *Journal of Medical Ethics* 4 at 6.
[17] Hurd, H.M. 'The Moral Magic of Consent' (1996) 2 *Legal Theory* 121
[18] Alexander, L 'The Moral Magic of Consent (II)' (1996) 2 *Legal Theory* 165.

This is the case regardless of whether the other person is, or is not, consenting. Thus, it is arguable that communication is integral to consent and some have gone so far as to suggest that: 'consent is performative rather than attitudinal'.[19]

These two approaches to the essential nature of consent perhaps reflect different political views of the atomistic-community autonomy debate. Seeing consent as an act of will draws comparison with Kant's view of autonomy, and denying or underplaying the role of communication accords with the Liberal view of autonomy as rational individualistic self-determination. This view might hold that, provided the other's autonomy has not been infringed the actor's behaviour was not wrong and, since the person is consenting (has exercised his autonomous will), it is irrelevant that the consent was not communicated. This extreme view of consent is insensitive to the actor's intention and is unable to distinguish the reckless actor from the one who cares. If this is placed in the context of healthcare, with its focus on beneficence and the patient's interests, this deficiency is brought into stark relief. However, the alternative view, which sees consent as solely an act of communication, ignores or underplays the importance of individual autonomy by focusing on community-based views of behaviour.

A moderate approach, that sees consent as both a state of mind and an act of communication, balances the individual's autonomy against the needs of the community.[20] Thus, Sherwin argues that consent is a 'social act', but one whose 'object is to express a particular mental act'.[21] The importance of communication is arguably implicit in the etymology of the term. Consent derives from the conjunction of the Latin words con and sentire, which mean to feel together. The idea of togetherness suggests a closeness of the two parties that could not be achieved without some form of communication. Restricting consent to a state of mind isolates the parties rather than bringing them together. Thus, it seems that communication is an essential element of consent, but can the same be said of the claim that consent is a state of mind?

Consent may be restricted to an act of communication by the construction of formal rules that are taken as signifying that consent. This requires some form of community or societal structure and theoretically does not require any particular state of mind since the rules can be constructed so as to avoid any need to enquire into the attitude of the other. For example, the rule could be that one consents to

---

[19] Wertheimer, A. 'Consent and Sexual Relations' (1996) 2 *Legal Theory* 89 at 94.

[20] A continuum may be visualized with the weight given to the person's state of mind at one end and the weight given to the act of communication at the other. Where that balance rests will primarily depend on one's conception of autonomy as more or less atomistic, or more or less socially situated.

[21] Sherwin, E. 'Infelicitous Sex' (1996) 2 *Legal Theory* 209 at 209, 217.

sex simply by accepting a red rose from the other person. If that were the case, it would not matter whether or not the person is willing as the signifying act of accepting the rose would be sufficient to make the act permissible. However, if – as I will argue later – the need for consent is predicated on a respect for the individual's autonomy, then the question of that person's willingness is crucial. Thus, it seems that both communication and the person's state of mind are necessary, or core, attributes of consent.

The implication of seeing consent as a state of mind is that the individual must know what it is that he is consenting to. Communication is also central to consent and this requires that consent be situated within the context of a relationship that allows the two parties to engage in the necessary discourse. However, communication is imperfect and when constructing the rules of consent this imperfection should be acknowledged. This is both legitimate and necessary: unlike natural kind concepts, consent is entirely socially constructed. In the absence of a society with rules of behaviour the concept would never have come into existence and the particular conception of consent that develops depends on the nature and values of that society. This will hopefully become apparent over the remainder of this article.

In the Oxford English Dictionary, consent – as a noun – is defined as: 'voluntary agreement to or acquiescence in what another proposes or desires; compliance, concurrence, permission'. While an agreement seems to express the mutuality inherent in the word's origins, it raises certain problems for the patient that may be undesirable. Margaret Gilbert suggests that: 'an agreement is, in effect, a joint decision', which entails a 'joint acceptance' of the agreed action(s), and, therefore, it 'requires... a "joint commitment"'.[22] This means that both parties are mutually obliged to honour the agreement and neither party may withdraw unilaterally.[23] For the patient this would mean that, in the absence of the doctor's permission, once he had consented to a procedure he would be unable to withdraw that consent without incurring some penalty. It was a reliance on this sense of consent, which may be appropriate to its use as a term in the context of formal contracts, that led Habiba to conclude that 'although the notion may appear to safeguard liberty, in fact once consent is given it entails binding obligations'.[24] However, given the beneficent nature of medical practice and the importance of

---

[22] Gilbert, M. "Agreements, Coercion, and Obligation" (1993) 103 (4) *Ethics* 679 at 691.
[23] Gilbert, M. "Agreements, Coercion, and Obligation" (1993) 103 (4) *Ethics* 679 at 693.
[24] Habiba, M.A. 'Examining consent within the patient-doctor relationship' (2000) 26 *Journal of Medical Ethics*, 183 at 184.

one's body to self-identity and autonomy, it seems inequitable to oblige the patient to stand by his consent in this way.

Gillon states that consent as a simple agreement 'is not relevant to medical interventions'.[25] Instead, 'consent means a voluntary, uncoerced decision, made by a sufficiently competent or autonomous person on the basis of adequate information and deliberation, to accept rather than reject some proposed course of action that will affect him or her'.[26] The problem with seeing consent as an 'acceptance' of a proposal is that it fails to distinguish between the propositions I willingly accept and those that I accept because I see them as inevitable and beyond my control. If the patient is seen as having a degree of control over what happens to him then it may be more appropriate for consent, at least in this context, to function as a form of permission rather than as an acceptance (or, for similar reasons, acquiescence, compliance or concurrence). The one caveat here is that, because 'permitting something' can mean a passive submission, which does not accord well with the autonomous self-determination, it is important that the mechanism of consent is seen as the more active 'giving permission'.

Fitting with its two main senses - as an agreement or as a permission - a valid consent either creates mutual obligations or it legitimates an otherwise illegitimate act. As I have already argued, it is consent as a permission that is most relevant in the context of consent to medical intervention. In legitimating an act, 'consent functions as a "moral transformative" by altering the obligations and permissions that determine the rightness of others' actions'.[27] Consent can morally transform an act and make it permissible only if the wrong breaches a right that I control. For example, if I own a bicycle I can give you permission to use it. However, if it were my brother's bike my consent would not affect the morality of your act in using the bike without his permission. Thus, the permission granted by my consent is morally transformative because it waives an underlying right that I control. The 'right' of consent is derivative on a more basic right and is parasitic on a societal structure that allows property, ownership (in this context, self-ownership) and justice.

When discussing consent to sex, Hurd suggests that: 'when we give consent, we create rights for others'.[28] This claim may be true for some instances of consent, such as when I consent to a contract, but in the context of medical treatment it is over inclusive. Consider the consent given to a doctor to perform an

---

[25] Gillon, R. *Philosophical Medical Ethics* (1985, 1996) Chichester: John Wiley & Sons at 113.
[26] Gillon, R. *Philosophical Medical Ethics* (1985, 1996) Chichester: John Wiley & Sons at 113.
[27] Alexander, L "The Moral Magic of Consent (II) (1996) 2 *Legal Theory* 165.
[28] Hurd, H.M. "The Moral Magic of Consent" (1996) 2 *Legal Theory* 121.

operation. The patient retains sufficient control of his right to bodily integrity to withdraw consent at any point.[29] Since control of a right normally follows the right, if the doctor had been given a right to operate the patient would have ceded control and would not be able to withdraw his consent. It may be possible to argue that the right has been granted but the control retained by the patient. However, it is perhaps more appropriate to speak of consent as generating permissions rather than rights per se.[30] In this context consent operates as a form of waiver rather than as a transfer of a right.[31]

Before moving on to the second part of this article, which will look at the moral basis for consent, I want to look at three claims that are sometimes made. First, it may be argued that the patient's right to consent carries with it a right to chose.[32] However, if consent works as a waiver that grants the doctor permission to act then it appears to be a negative liberty right, which means that the only choice arising directly from the right to consent is that of giving or withholding consent to the treatment on offer. This does not seem to be much of a choice, especially if the patient feels that he has little option but to accept some form of medical treatment. However when combined with the doctor's legal duty of care (in negligence), or his moral duty of beneficence, this negative liberty right is given new bite.

Assume that the patient suffers from a condition that is amenable to a range of treatments and, although some are more efficacious than others, all remain medically indicated. Because it is the treatment that he had most experience with, the doctor offers the patient treatment A. Treatment A has a side effect that the patient particularly wants to avoid so he refuses consent. However, even though he has refused consent to a particular treatment he remains the doctor's patient and the duty of care/beneficence persists. Since the refusal of consent, by a competent adult, acts as a veto, that treatment is removed from the doctor's

---

[29] Once an operation is underway it may be pragmatically difficult to completely withdraw consent, but the patient is free to vary his consent and limit the doctor to repairing any damage caused by the operation, such as stopping any bleeding and closing the incision.

[30] McConnell, T. *Inalienable Rights: The Limits of Consent in Medicine and the Law* (2000) New York: Oxford University Press at 8.

[31] See: McConnell, T. *Inalienable Rights: The Limits of Consent in Medicine and the Law* (2000) New York: Oxford University Press. See also: Sherwin, E. "Infelicitous Sex" (1996) 2 *Legal Theory* 209 at 217.

[32] Austoker, J. "Gaining informed consent for screening" (1999) 319 *British Medical Journal* 722; Worthington, R. "Clinical issues on consent: some philosophical concerns" (2002) 28 *Journal of Medical Ethics* 377, at 378; Capron, A.M. "Informed Consent in Catastrophic Disease Research and Treatment" (1974) 123 *University of Pennsylvania Law Review* 340, at 349.

armoury,[33] but the patient's illness still needs treating and this remains the doctor's responsibility. In some settings, the doctor might suggest that the patient see another doctor and some degree of choice of treatment arises from the choice of doctors. However, assume for the sake of argument that, for one reason or another (the patient may already have experienced the same scenario with all the other reasonably accessible doctors), the option of going to another doctor is closed. The only option for the doctor in this situation is to suggest another treatment or abandon the patient, which may breach both his legal and moral duty.

Treatment B is the next best treatment in terms of clinical efficacy. However, again the patient refuses consent, this time because of one of the risks associated with the treatment. This leaves the doctor with the option of offering treatment C or abandoning the patient. Again his legal duty of care, and certainly his moral duty of beneficence, requires him to offer treatment C, even though it is much less clinically effective. The difficulty with this scenario is that the patient may be unaware of the other treatment options and so may consent to treatment A out of fear. However, the doctor's duty to inform the patient of alternative treatments, which is required by his obligation to obtain the patient's informed consent, should allow him to be alerted to these options. This might be countered by suggesting that the duty to disclose alternatives only extends to those available to the patient and if the doctor is unwilling to offer a treatment it does not need to be disclosed. However, the argument explicated above suggests that all medically indicated treatments should be seen as options *ab initio*, despite the doctor's reluctance. Thus, the combined power arising from veto of refusing consent and the doctor's duty of care/beneficence indirectly entail a choice of medically indicated treatment options.

Although consent may play a role in patient choice, it is important to remember that, in the absence of the doctor's duty of care/beneficence, consent is neutral. However, the doctor's duty to respect the patient's autonomy adds weight to the argument that those choices should be offered to the patient without making him go through the rigmarole of refusing treatments until an acceptable one is offered. The caveat to this is that, because consent acts only in synergy with other moral obligations it is unhelpful, and perhaps disingenuous, to try and tag positive claims on to consent's coat tails. To do so circumvents the necessary arguments of justice that ground an equitable distribution of resources. Consent is not a tool for consumerist claims,[34] and whether or not a treatment should be available is more

---

[33] This may not be the case for some mental illnesses and some communicative diseases.

[34] Alderson, P. Goodey, C. "Theories of consent" (1998) 317 BMJ 1313: the authors distinguish 'choice' as one of the five models of how consent is used.

properly determined by arguments of clinical efficacy, cost, the goals of medicine, and professional autonomy balanced by the doctor's duty of beneficence and the patient's right to positive autonomy.

A second claim that I will briefly consider is that consent is a process.[35] There are two advantages to seeing consent this way. First, it reflects the reality that, in many cases, disclosure, discussion, and negotiation between doctor and patient extends over time and it supports the idea of shared decision-making (see below).[36] Second, it reminds the doctor that treatment may be prolonged and that he should ensure that the patient is still consenting throughout. The problem, however, is that it concatenates consent with the process that leads up to it. As I have already argued, consent is a state of mind that is crucially dependent on an act of communication. Consent cannot be both a process and a state of mind and if it is to function as a permission (or waiver) that legitimates an action then it must be seen as an event: there must be a point in time when the act is illegitimate and a subsequent point – following consent – when it is permissible. The concept of a process is too diffuse to deal with the discrete change in the status of the act.

The third claim is intimately tied up with the idea of consent as a process. It argues that consent is a shared decision. Again this has its merits as it supports the idea of the doctor and patient working as cooperative partners in a mutual relationship. However, it cannot be the case since only the patient – as a competent adult – has the right to give or withhold consent. Even if that patient cedes the treatment decision entirely to the doctor, he must do so by giving his consent. There may well be shared decision-making prior to consent but the competent patient cannot avoid the decision to give or withhold consent. As Beauchamp and Childress have argued, shared decision-making is a 'worthy ideal in medicine' but one that 'neither defines nor displaces informed consent'.[37] My approach may be criticised as overly formal and unsympathetic to the patient's needs. It is not intended to be seen that way and it is formal only in the sense that I believe it is helpful to distinguish the core attributes, which are those justified by the theory that supports the concept's existence, from those attributes that are not

---

[35] Usher, K. J. Arthur, D. "Process consent: a model for enhancing informed consent in mental health nursing" (1998) 27 *Journal of Advanced Nursing* 692; Dyer, A.R. Bloch, S. "Informed consent and the psychiatric patient" (1987) 13 *Journal of Medical Ethics* 12; Kay, R. Siriwardena, A.K. "The process of informed consent for urgent abdominal surgery" (2001) 27 *Journal of Medical Ethics* 157.

[36] Meisel, A. Kuczewski, M. "Legal and Ethical Myths about Informed Consent" (1996) 156 *Archives of Internal Medicine* 2521 at 2522.

[37] Beauchamp, T.L. Childress, J.F. *Principles of Biomedical Ethics*, 5th edition. (2001) New York: Oxford University Press at 78.

central to the concept and should, where the concept is socially constructed, be justified independently.

The patient's right to give or withhold permission to an intervention is readily justified by a Liberal or Libertarian politic. Since consent is based on a negative liberty right of non-interference it is a strong position and few people in a democratic society would suggest that consent in this sense was unnecessary. [38] Because an accepted negative liberty right provides a strong base, there is a temptation to try and use it to make positive rights claims. However, because these claims require the provision of resources they raise very different issues about the nature of healthcare, the role of the state and the relationship between the individual and the state. Rather than hiding these issues behind a social concept that protects the negative liberty of an individual, it would be better to justify these other claims as goods independent of, but subject to, consent. Furthermore, shared decision-making, negotiation, mutual trust, discourse and disclosure are all goods, irrespective of whether consent is required. To keep them independent of consent makes it easier to see them as necessarily pervasive rather than restricted to treatment events.

As an example, consider the decision to issue a 'do not resuscitate' (DNR) order. The decision to make a DNR order should be based on whether resuscitation would be clinically efficacious. The decision may also take into account resources and funding, but the primary basis should be whether there would be a reasonable chance of benefiting the patient. This is essentially a clinical decision, although one that should consider the patient's perspective on what counts as a benefit. Before the right to give or withhold consent is relevant, the patient must show that he has a right to be offered the treatment, as it is only if he is offered the treatment that his consent is required. Unless resuscitation is clinically indicated the doctor is under no obligation to provide it and therefore the patient has no right to be offered it. Any claim to treatment must be based on an appeal to some objective of distributive justice, such as capacity to benefit. But this claim is of a different sort to the right of consent and is logically prior to that claim. It would be better to see the decision to issue a DNR order as one that

---

[38] It has been argued that the patient's consent is 'redundant' because the patient has 'requested' treatment. Space does not permit a complete rebuttal of this argument. However, a request, if it has any effect, acts by subsuming consent not by making it irrelevant: consent is implicit in any request. One caveat is that should the doctor wish to vary from that request then an additional consent would be required. A second caveat is that the patient may make a very general request for treatment because he does not know what is wrong. This again requires the doctor to seek additional consents for specific interventions since the patient is unlikely to have had them in mind when requesting the doctor's help. See: Habiba, M.A. "Examining consent within the patient-doctor relationship' (2000) 26 *Journal of Medical Ethics* 183.

should be made following discussion and with the patient's agreement rather than the stronger claim that it requires his consent: the patient has a right to be involved in discussion and a right to permit or veto resuscitation, but has no right to resuscitation attempts *per se*.[39]

THE MORAL BASIS OF CONSENT

Personal autonomy is widely, if not universally, acknowledged as the moral basis for consent.[40] It is, however, based only on the negative aspect of autonomy. This is not to diminish the importance of positive autonomy, merely to suggest that positive autonomy claims require independent justification. Thus, as I suggested earlier, consent cannot drive patient choice, which requires the additional duties of beneficence and/or a wider respect for autonomy. I do not intend to try and justify the importance of autonomy,[41] however, it is relevant to note that there are many different conceptions of autonomy, which remains a contested concept.[42] These different conceptions are driven by competing political views concerning the nature of society and the context of the individual within the community. They can be broadly divided into three groups. The first group essentially comprises autonomy as independence or self-determination,[43] which is the most literal interpretation of the term's etymological origins: autonomy is derived from Greek and means self rule.[44] The second group sees autonomy as requiring the capacity for rational or reflective self-determination. One well-known inception is the requirement that the individual of assesses first order desires on the basis of more stable second order desires.[45] The third group requires

---

[39] An argument may be made in support of a claim right to resuscitation, but only by bringing in to play a 'right to life' that imposes appropriate positive obligations on the state and the method of resuscitation is not physiologically futile.

[40] Worthington, R. "Clinical issues on consent: some philosophical concerns" (2002) 28 *Journal of Medical Ethics* 377 at 377.

[41] See, for example, Hurka, T. "Why Value Autonomy?" (1987) 13(3) *Social Theory and Practice* 361.

[42] Dworkin, G. *The Theory and Practice of Autonomy* (1988) Cambridge: Cambridge University Press at 9.

[43] Benson, J. "Who is the autonomous man?" (1983) 58 *Philosophy* 5 at 8-9.

[44] Dworkin, G. *The Theory and Practice of Autonomy* (1988) Cambridge: Cambridge University Press at 12.

[45] Dworkin, G. *The Theory and Practice of Autonomy* (1988) Cambridge: Cambridge University Press at 20. The idea of second order desires and, more appropriately, second order volitions, was first proposed as a way of distinguishing persons from non-persons by Harry Frankfurt in: "Freedom of the Will and the Concept of a Person" reprinted in: Kane, R. (ed.) *Free Will* (2002)

autonomy to be both rational and moral. Kant's view of autonomy of the will as permitting only self-legislation that is capable of universal application is one example,[46] and the concept of relational autonomy is another.[47]

The relevance of these different conceptions of autonomy is that the conception of consent will vary according to which model of autonomy is adopted. The point is not to argue in favour of one view over another but to indicate the variables and values these different conceptions of autonomy import into consent. One of these is the degree to which autonomy is socially embedded. This might be seen on a continuum with an extreme Libertarian view of the atomistic individual at one end with an extreme communitarian view of the social self, in which, if autonomy is a relevant concept at all, the individual is both constituted by and dependent on his relations with the community and other individuals within it. Lying in between these two extremes would be the Liberal view, which recognises the importance of social relations but still sees individual identity as determined by one's autonomous choices.

One consequence of this variable is that the more socially embedded view justifies a greater degree of constraint on individual self-determination as the individual is not acting autonomously unless his decision pays due consideration to the relational others. Thus, the range of interventions controlled by the individual's consent may be more restricted and procedural requirements to consider others in one's decision may be justified. A second potential consequence is that the patient is not obliged to be an isolated decision-maker and reliance on the support of others can be justified by the relational dependence of autonomy. A third consequence is the recognition that, because autonomy is necessarily relational, attention should be given to *both* the immediate negotiation between doctor and patient *and* the surrounding 'practices and policies' that contextualise the relationship and influence the patient's autonomy.[48] A fourth consequence is that consent, which is justified by negative autonomy, is of limited

---

Malden (Mass): Blackwell Publishers 127-144. See also: Richards, D.A.J. "Rights and Autonomy" (1981) 92(1) *Ethics* 3 at 13.

[46] Kant, I. *Groundwork of the Metaphysics of Morals* (Gregor, M. *Transl.*) (1998) Cambridge: Cambridge University Press at 47 (4:440).

[47] Mackenzie, C. Stoljar, N. (eds) *Relational Autonomy: Feminist Perspectives on Autonomy, Agency and the Social Self* (2000) New York: Oxford University Press.

[48] Dodds, S. 'Choice and Control in Feminist Bioethics' in Mackenzie, C. Stoljar, N. (eds) *Relational Autonomy: Feminist Perspectives on Autonomy, Agency and the Social Self* (2000) New York: Oxford University Press 213 at 215.

value in empowering patients and that the debate about the implications of respecting autonomy does not start and finish with a discussion of consent.[49]

The second relevant variable is that of rationality. Again this may be seen as a continuum with no requirement for rationality at one end of the extreme to maximal or perfect rationality at the other. The obvious consequence of this is on the determination of whether the individual possesses sufficient capacity for autonomy to be deemed competent. Where no rationality is required, the individual need only be capable of making a choice, which is a very minimal requirement and risks abandoning the more vulnerable patients. Too great a requirement for rationality, however, may limit the population of competent individuals to the privileged few whose values, views and decisions accord with those in authority.

The issue of how much rationality should be required is complicated by the differing conceptions of rationality that may be employed. Models of rationality may be split into two types. In the recognitional model, rationality is measured against a list of objective goods, which may in turn be hierarchically ordered so that it would be seen as irrational to choose a lower order good over one that has a higher objective value.[50] The constructivist model may be divided into two variants. In the first type, the neo-Humean, rationality is measured against the individual's subjective goods.[51] These goods are not simply the agent's desires but are arrived at through knowledgeable and reasonable reflection.[52] The second constructivist approach is the Kantian model that equates morality and rationality and determines the question by reference to Kant's Imperative of universalisability.[53] Each of these views has its strengths and weaknesses and all have supporters and detractors. Again, the point in raising this here is not to argue in favour of one or other view but simply to highlight some of the issues that may

---

[49] Dodds, S. 'Choice and Control in Feminist Bioethics' in Mackenzie, C. Stoljar, N. (eds) *Relational Autonomy: Feminist Perspectives on Autonomy, Agency and the Social Self* (2000) New York: Oxford University Press 213 at 213.

[50] Gaut, B. 'The Structure of Practical Reason' in Cullity, G. Gaut, B. (eds.) *Ethics and Practical Reason* (1997) Oxford: Clarendon Press 161 at 161-162.

[51] Cullity, G. Gaut, B. 'Introduction' in Cullity, G. Gaut, B. (eds.) Ethics and Practical Reason (1997) Oxford: Clarendon Press 1 at 4.

[52] Cullity, G. Gaut, B. 'Introduction' in Cullity, G. Gaut, B. (eds.) *Ethics and Practical Reason* (1997) Oxford: Clarendon Press 1 at 7-8.

[53] Cullity, G. Gaut, B. 'Introduction' in Cullity, G. Gaut, B. (eds.) Ethics and Practical Reason (1997) Oxford: Clarendon Press 1 at 3-5; Kennett, J. *Agency and Responsibility: A common-sense moral psychology* (2001) Oxford: Clarendon Press, at 98; Koorsgard, C.M. 'The Normativity of Instrumental Reason' in Cullity, G. Gaut, B. (eds.) *Ethics and Practical Reason* (1997) Oxford: Clarendon Press 215 at 231.

lie hidden behind an appeal to autonomy as the moral justification for respecting someone's right to give or withhold consent.

Before going on to discuss the issue of consent and responsibility, there is an additional distinction to note. Unless one adopts the conception of autonomy as straightforward self-determination, then the autonomous person may be distinguished from the autonomous act. Just because someone has the capacity to act rationally and/or morally does not mean that his every act will be rational and/or moral. The question is then whether a respect for his autonomy requires protection of any decision or only those that are in themselves autonomous. On the basis of this distinction, interfering with someone's non-autonomous decision is an infringement of his liberty. However, if one believes that autonomy has a greater value than liberty this may seem acceptable.[54] The harm done in infringing liberty may be more than offset by the protection of that person's future autonomy. Thus, an autonomist may place a limit on the right to give or withhold consent so that decisions that would be foreseeably catastrophic could be justifiably overridden. Such a view would justify the use of a competence test based on a 'reasonable outcome' of choice. Roth et al's approach to this test is illuminating. They state: 'The benefits and costs of this test are that social goals and individual health are promoted at considerable expense to personal autonomy'.[55] This conclusion, however, depends on either seeing autonomy as self-determination or prioritising liberty over autonomy. It also, crucially, depends on which model of rationality one adopts as the neo-Humean may reach a very different opinion to the Kantian or those who rely on the recognitional model.

RESPONSIBILITY FOR OUTCOME

In this section I will look at the question of whether responsibility for outcome follows consent: when I consent to an operation I grant the surgeon permission but do I also agree to bear responsibility for all of the consequences? It might be thought that if I am a competent agent and I voluntarily agree to something then I should also accept the consequences of that decision. However, wherever consent is granted there must also be at least one other competent agent involved in what has become a joint venture. When placed in a social context there are three parties who might bear responsibility for outcome: the patient; the

---

[54] Haworth, L. *Autonomy: An Essay in Philosophical Psychology and Ethics* (1986) New Haven: Yale University Press at 131.
[55] Roth, LH. Meisel, A. Lidz, CW. 'Tests of Competency to Consent to Treatment' (1977) 134 (3) *American Journal of Psychiatry* 279.

professional; and the community or state. In a Liberal or Libertarian system, the loss is allowed to lie where it falls, unless it was caused by wrongdoing. The work that consent does here is to prevent the intervention from being wrong and while this indirectly determines the allocation of responsibility it is really down to justice rather than consent. Where it is required to legitimise an action then the individual's consent is necessary to allow him to be included in the distribution of responsibility. It is not, however, sufficient to determine that distribution.

In a Libertarian system, responsibility for outcome may be determined entirely by negotiation between the two 'free' and equal contracting parties. However, as soon as it is recognised that the two parties are rarely, if ever, equal negotiators and that the circumstances that provide the context to the negotiation may undermine the perception of freedom, it is arguable that factors other than consent are relevant to determining responsibility. The situation is made more complex as it is almost certain that the agreed terms will underdetermine responsibility. The Libertarian response might be that where there is no specific agreement and there has been no wrongdoing then the loss should lie where it falls. This is, however, but one way of allocating responsibility and other approaches are possible. This simple Libertarian argument is predicated on the idea of corrective justice, which essentially requires a wrongdoer to make good any loss he has caused the innocent party. Corrective, or rectificatory, justice has long been an accepted principle of justice and was described by Aristotle, along with distributive justice, as one of the two basic types of justice.[56]

For Aristotle, distributive justice concerns the fair allocation of divisible resources while rectificatory justice aims to 'equalise' the losses and gains arising from an unfair transaction.[57] Distributive justice operates at the level of the community while corrective justice is engaged by interactions between individual parties and some authors, such as Weinrib, argue that they 'are structurally different and mutually irreducible... [and] cannot be assimilated to each other'.[58] If this is the case then an area of law, like tort, that resolves private conflicts between individuals should be based purely on the principle of corrective justice.[59] However, in some recent English and Scottish legal cases, the House of Lords

---

[56] Aristotle. *Ethics* V. ii-iii 1130b-1132a (Thomson, J.A.K. transl.) (1976) London: Penguin at 176-180.
[57] Aristotle. *Ethics* V. ii-iii 1130b-1132a (Thomson, J.A.K. transl.) (1976) London: Penguin at 176-180.
[58] Weinrib, E.J. 'Legal Formalism: On the Immanent Rationality of Law' (1988) 97 *Yale Law Journal* 949 at 983.
[59] Wright, R.W. 'Right, Justice and Tort Law' in: Owen, D.G. (ed.) (1995) *Philosophical Foundations of Tort Law*, Oxford: Clarendon Press 159 at 172.

made explicit appeals to distributive justice in order to limit or exclude liability.[60] In *McFarlane v Tayside Health Board*, Lord Steyn explained that tort law is: 'a mosaic in which the principles of corrective justice and distributive justice are interwoven'.[61] If this is the case then there must be a way of connecting the two principles.

One possible way of rationalising distributive and corrective justice is to adopt an animated rather than static model of society.[62] The difference between the two is that, unlike the static model, which allocates resources on the basis of distinct, non-interacting individuals, the animated model acknowledges the inevitability of private interactions that will impact on any initial distribution. In taking account of these interactions, the animated model is able to recognise both the positive and negative resources. The positive resources are those goods that are normally associated with distributive justice: the tangible goods such as food, money and property, and the intangible goods such as liberty. An additional positive resource is the benefit of a favourable transaction. However, just as a transaction may result in gain so it may also cause loss. While there are transactions in which all parties gain, there are also those where one party stands to gain at the expense of the other who suffers a net loss. These losses are the negative resources that are also subject to the principles distributive justice. In determining responsibility for the losses and gains, the Libertarian approach, perhaps exemplified by Honoré's theory of 'outcome-responsibility',[63] is but one mode of distribution. In other words, corrective justice is simply one of many possible approaches to ensuring that losses and gains are fairly distributed. It has no special standing and is as open to challenge as any other theory, or part of a theory, of distributive justice.

What this means for consent is that, as I suggested earlier, the presence of consent is only one factor that should be taken into account when determining who should be responsible for any losses caused by the intervention. While the

---

[60] *White v Chief Constable of South Yorkshire Police* [1999] 2 AC 455; *Rees v Darlington Memorial Hospital NHS Trust* [2003] UKHL 52. For a discussion of the role of distributive justice in these cases see: Maclean, A.R. 'Distributing the Burden of a Blessing' (2004) I *Journal of Obligation and Remedies* 23.

[61] *McFarlane v Tayside Health Board* [2000] 2 AC 59 at 83.

[62] Maclean, A.R. 'Distributing the Burden of a Blessing' (2004) I *Journal of Obligation and Remedies* 23 at 26.

[63] Honoré, T. 'The Morality of Tort Law – Questions and Answers' in: Owen, D.G. (ed.) (1995) *Philosophical Foundations of Tort Law*, Oxford: Clarendon Press 73 at 81. Simply put, it is the view that if one acts and seeks to claim the benefits one should also be responsible for any harms or losses that result. See also: Lippke, R.L. 'Torts, Corrective Justice, and Distributive Justice' (1999) 5 *Legal Theory* 149 at 153.

Libertarian would treat consent as the sole consideration, recognising the relevance of distributive justice and openly considering these concerns may offset the harshness of that approch, which pays no heed to suffering, need or welfare. Such issues were apparent in the recent series of wrongful pregnancy[64] cases that came before the English courts.[65] Although the relevant cases did not involve consent issues,[66] wrongful pregnancy claims may arise from a failure to disclose the risk of failure of sterilisation and thus the arguments are applicable.

In *McFarlane*, the House of Lords held that the maintenance costs for raising a healthy child would not be recoverable even though they would have been on principles of corrective justice. Although the distributive justice arguments were not made clear, it is arguable their Lordships relied on *desert* as a limiting principle.[67] In *Parkinson*, however, the child was disabled and here the Court of Appeal distinguished *McFarlane* and, relying on the additional *needs* of the family raising a disabled child as the limiting principle, held that the extra costs of raising the child could be recovered. A similar decision was made in the Court of Appeal hearing of *Rees*: although it was the mother who was disabled rather than the child, the Court of Appeal still appealed to her *need* as justifying recovery of the additional costs.[68] One of the problems with that decision, which was recognised by the dissenting judge, was that the triggering factor was the presence of a physical disability rather than the claimant's need per se. Thus, recovery would be denied where the *need* was due to the social circumstances of the mother, even if the additional stress caused by having to care for the new child would have subsequently caused the mother to become ill. It was, at least in part, because of the arbitrariness of this that the House of Lords, by a 4:3 majority, overturned the decision (although the majority softened the harshness of its decision by awarding a conventional sum for the harm done to the woman's autonomy).

---

[64] Wrongful pregnancy, or wrongful conception, claims are brought when the professional's negligence is responsible for the claimant's conceiving a child, which is what they had sought to avoid by being sterilized. The claims are both for the harm of pregnancy and the maintenance costs involved in raising the child.

[65] *McFarlane v Tayside Health Board* was a Scottish case, but the relevant hearing was before the House of Lords, which is the highest court of appeal for civil cases in both England and Scotland.

[66] *McFarlane v Tayside Health Board* involved negligent advice; *Parkinson v St James and Seacroft University Hospital* [2001] 3 All ER 97 and *Rees v Darlington Memorial Hospital NHS Trust* [2003] UKHL 52 involved negligent performance.

[67] Maclean, A.R. 'Distributing the Burden of a Blessing' (2004) 1 *Journal of Obligation and Remedies* 23 at 27.

[68] *Rees v Darlington Memorial Hospital NHS Trust* [2002] EWCA Civ 88.

These cases perhaps illustrate two things. First, the split in the judges' decisions, and the attempts by the Court of Appeal to distinguish *McFarlane*, suggests that the courts are not the best place to be making this kind of distributive justice assessments, which are really a political matter for the Executive and Legislature to determine. More importantly for present purposes, however, the open appeal to principles of distributive justice is acknowledgment of the harsh injustice of relying purely on corrective justice. Rather than making one or other party wholly responsible for the loss, it may be better to distribute the losses more evenly between the parties. A number of factors may be brought into the equation. For example, the allocation of loss may be affected by the parties' behaviour: the greater the wrongdoing the greater the share of the burden. The inequality between the two parties, of power and knowledge, ought to be considered in determining responsibility. It may also be relevant to take into account which party stood to benefit from the interaction and to what extent.[69] Other factors, such as the parties' needs, or the utility of the allocation may also be considered.

The point of this discussion is not to argue for any particular distribution of responsibility, but merely to show that consent should not be seen as determinative. There are factors, other than agency, that are relevant to this and it is important to recognise the political assumptions that affect the association between consent and responsibility for outcome.[70] It is also important to acknowledge that, in a community that provides welfare support, such as free healthcare, there is already a basic level of shared responsibility for losses and burdens. However, these welfare provisions are variable from community to community and should not be relied on to justify creating a necessary link between consent and responsibility for outcome. As a final illustration of why consent and responsibility for outcome should be less strongly associated, I will briefly discuss a recent case that came before the House of Lords.

In *Chester v Afshar*, a 51-year-old female journalist was left with a serious neurological deficit following a competently performed operation to relieve her back pain. The surgeon had failed to disclose that the operation carried a risk of nerve damage or paralysis and the court held that he had breached his duty of care to the patient. However, in order to make the surgeon liable, the claimant had to

---

[69] See, for example, Paragraph 15, Declaration of Helsinki 1964 (as amended), which holds the researcher (who stands to benefit from the research) responsible for any consequences irrespective of the consent of the subject. Available at: http://www.wma.net/e/policy/b3.htm (accessed 14th December 2004)

[70] Context may also be important in that different factors are likely to be relevant when distributing responsibility in, for example, sexual relationships, business relationships or political relationships.

show that this failure to warn her of the risk was causally responsible for the damage. This requirement reflects the principle of corrective justice in a relatively pure form and traditionally it has been satisfied only when the claimant has been able to satisfy the court that, had she been warned of the risk, she would have withheld consent to the procedure and opted for a different treatment. In the present case, Ms Chester was only able to say that she would have sought a second or third opinion. She was unable to say that she would not have subsequently consented to the same operation. The difficulties in this case were that the risk was small (1-2%) but the harm was serious; the operation was performed competently and the failure to disclose may be seen as an easy mistake to make – an oversight that just reflects human fallibility rather than any intention to manipulate the patient's decision; even had the risk been disclosed she would – in all likelihood – have agreed to the operation at some point; although the surgeon would gain financially (it was a private operation), the procedure was primarily for the patient's benefit; and, although the surgeon's error may be seen as a minor transgression, the damages would be very high.

The House of Lords was split. The majority followed the Court of Appeal and found in the claimant's favour. Their decision, which was recognised to be an extension of the traditional rules of causation, was justified by an appeal to the importance of protecting the patient's autonomy.[71] However, this justification for holding the surgeon responsible is persuasive only if the more individualistic Liberal or Libertarian approaches to autonomy are adopted. In these views, consent is akin to a contract between two equal competitors, the self is constructed through his choices and responsibility requires authorship. However, if a more socially embedded approach to autonomy is taken then factors other than the individual's decision become relevant to the ascription of responsibility. For example, recognising that power imbalances may affect decision-making, even in the absence of coercion,[72] may allow a more sensitive approach that does not necessarily require one or other party to bear full responsibility. Furthermore, a more social approach allows consideration of the community's responsibility and the impact on the community of extending liability in this case.

The two Lords in the minority did not explicitly appeal to a more social conception of autonomy. Instead they effectively argued that materialisation of the risk was simply a matter of bad luck, and, as she would have faced the same risk on another occasion, it would be unfair to hold the surgeon responsible for his

---

[71] *Chester v Afshar* [2004] UKHL 41 [24].

[72] Clement, G. Care, *Autonomy, and Justice: Feminism and the Ethic of Care* (1998) Boulder (Col): Westview Press at 25.

oversight. As Lord Bingham concluded: 'The patient's right to be appropriately warned is an important right, which few doctors in the current legal and social climate would consciously or deliberately violate. I do not for my part think that the law should seek to reinforce that right by providing for the payment of potentially very large damages by a defendant whose violation of that right is not shown to have worsened the physical condition of the claimant'.[73] Neither of the two Lords in the minority provided strong explanations for their judgments. However, it is possible to justify their approach by taking into account the potential effect on the community.

Although this case involved a private patient, English courts have generally held that the doctor's duty is the same whether the case involved a private or an NHS patient.[74] There is a belief that the NHS is facing a litigation crisis and allowing recovery in this case would add fuel to the fire.[75] Although this particular case would not drain the NHS coffers, it would open the door to allow other cases that might do so. If the NHS has to meet a large litigation bill, there will be less money for other patients, who will suffer as a consequence and this may, in future, also include the claimant herself. A second effect of adding to this perceived litigation crisis, is that it may deter some from entering into the profession, or encourage others to leave or enter low risk specialities. This, again, would be detrimental to all, including the claimant. Furthermore, the NHS will provide essentially free healthcare to the claimant, which will lessen the impact of her injury. As well as that support, she will be able to get other Government funded help. These community provisions mean that at least some of the financial responsibility for the injury is shared.

The ideal resolution would have been a more explicit, context sensitive, justified distribution. The courts, however, are constrained by precedent and a limited range of remedies such that one or other of the two parties must be held wholly responsible in law. Even though the party may be wholly legally responsible, the economic burden may be lightened by other mechanisms. If the

---

[73] *Chester v Afshar* [2004] UKHL 41 [9].

[74] See *Thake v Maurice* [1986] QB 644 at 679 per Kerr LJ, CA.

[75] V. Harpwood, 'The Manipulation of Medical Practice' in M. Freeman, A. Lewis, (eds.), *Law and Medicine: Current Legal Issues Volume 3* (2000) Oxford: Oxford University Press at 47, 48; L. Mulcahy, 'Threatening Behaviour? The Challenge Posed by Medical Negligence Claims' in M. Freeman, A. Lewis, (eds) *Law and Medicine: Current Legal Issues Volume 3* (2000) Oxford: Oxford University Press at 81, 83. Although Mulcahy commented that her research suggests the threat was exaggerated, in 2001, the National Audit Office (NAO) reported a seven-fold increase in costs since 1995-1996: NAO, *Handling Clinical Negligence Claims in England* (London 2001), 1 at: http://www.nao.gov.uk/publications/nao_reports/00-01/0001403.pdf (last visited 14th December 2004).

claimant is left with responsibility then, in communities that provide good welfare and healthcare support, the responsibility is shared to at least some extent. If the doctor is made responsible then the loss is shared either through malpractice insurance or through the NHS Indemnity Scheme. The sharing is limited to smaller communities within the state and is not particularly sensitive to context-specific factors. Furthermore, the danger of this approach is that it risks undervaluing the patient's autonomy. One possible response to this would be to acknowledge that the patient's autonomy might be harmed independently of the outcome. This might be achieved simply by awarding damages on the basis of the degree of infringement to the patient's autonomy. This possibility was perhaps recognised by Lord Hoffmann's argument that while the doctor should not be responsible for the full cost of the injury 'a modest solatium' might be paid because the 'failure to warn ... was an affront to her personality'.[76] Although the House of Lords in *Chester* did not follow that approach, a parallel may be drawn to the award of damages for harm to the claimant's autonomy in *Rees* (see above). By recognising infringement of autonomy as a distinct harm, the law is then free to make a more distributively fair allocation of responsibility for outcome, especially where luck plays a significant part.

CONCLUSION

In this article I have looked at three aspects of consent: its nature, its basis and its consequences. Space has prevented a comprehensive exploration and the primary aims were to reveal hidden or underlying assumptions and to examine the relationship between consent, autonomy and responsibility for outcome. From this brief discussion it is apparent that consent is essentially a political concept that is dependent on the relationship between the individual and the community. Both its basic structure and its meaning are shaped by the political currents that flow around consent's banks. The water is muddied further by the assumptions that eddy consent's bed of autonomy and rationality. The implication of these, often hidden, assumptions is that consent can support a wide range of obligations and rights. This is compounded by the tendency to use consent to shore up positive autonomy claims to choice or to influence other communicative aspects of the professional-patient relationship. It is, I believe, important to keep these assumptions and influences apparent. A failure to recognise the real motivations for certain claims may mean that more contentious obligations are accepted

---

[76] *Chester v Afshar* [2004] UKHL 41 [33-34].

because of the strength of the claim-right to control interference. Recognising the political influences that affect both the core concept and its more peripheral features allows any claims that derive from them to be visible and available for debate.

One aspect of consent that is perhaps rarely challenged is the association between consent and responsibility for outcome. Because of the reliance of consent on the underlying principle of autonomy, the link superficially seems a natural one. I argued, however, that the connection is based on Libertarian assumptions about autonomy, identity and responsibility. Adopting a more social or relational view of autonomy and consent allow the two concepts to be at least partially disconnected. If responsibility for outcome is seen as a negative resource, and the association with autonomy is weakened, other distributive objectives may be brought into the equation allowing a more responsive approach to responsibility that is sensitive to issues such as power imbalance, need, and luck. It is important, therefore, to be aware of these influences and assumptions that shape consent and the obligations and rights that flow from legal regulation.

In: Contemporary Ethical Issues
Editor: Albert G. Parkis, pp. 63-86
ISBN 1-59454-536-7
© 2006 Nova Science Publishers, Inc.

Chapter 4

HUMAN XENOTRANSPLANTATION: AN IMMUNOLOGICAL AND ETHICAL CHALLENGE

Y. T. Ghebremariam[1], S. A. Smith[1,2], J. B. Anderson[2], D. Kahn[3] and G. J. Kotwal[1,2]*

[1] Division of Medical Virology, IIDMM, University of Cape Town 7925, HSC, Cape Town, South Africa.
[2] Department of Microbiology and Immunology, University of Louisville School of Medicine, Louisville, Kentucky 40202, USA.
[3] Division of General Surgery, Department of Surgery, University of Cape Town, GSH, HSC, Cape Town, South Africa.

ABSTRACT

Despite the preliminary success in cellular and tissue xenotransplantation, the transplantation of solid organs across species has been a monstrous challenge in the history of xenotransplantation research. In principle, the transplantation of organs from taxonomically related concordant species has a survival advantage over the use of organs from unrelated discordant species. Yet, clinically, the survival of concordant xenografts has only been measured in days. Moreover, the use of organs from non-human primates may not be feasible due to logistic and ethical barriers. However, to benefit from non-primate organs, the biochemical and metabolic rifts created by the phylogenetic differences need to be resolved. Different haematological and genetic manipulations have been tried alone or in

---
* Girish J. Kotwal: gkotwal@curie.uct.ac.za

combination to prolong graft survival. Convergence of these strategies together with the administration of soluble complement control proteins, such as the vaccinia virus complement control protein (VCP), may provide a useful adjunct to prolong the survival of xenografts. However, a number of ethical and moral issues have obscured the topic of xenotransplantation and introduced yet another challenging barrier. This article will discuss the immunological and biopsychosocial issues involved in human xenotransplantation.

ABBREVIATION LIST

Accelerated rejection	AR
Acute cellular rejection	ACR
Acute vascular rejection	AVR
Alkaline Phosphatase	ALP
Antibody-dependent Cell-mediated Cytotoxicity	ADCC
Chronic rejection	CR
Cluster of differentiation 46, 55, 59	CD46, CD55, CD59
Cobra venom factor	CVF
Decay-accelerating factor	DAF
Delayed xenograft rejection	DXR
Endothelial cell	EC
Galactosyl	Gal
Gastrointestinal	GI
Heparan sulfate proteoglycan	HSP
Hyperacute rejection	HAR
Immunoglobulin G/M	IgG/IgM
Interleukin	IL
Macrophage inflammatory protein	MIP
Membrane cofactor protein	MCP
Natural killer cells	NK
Plasminogen activator inhibitor type-1	PAI-1
Platelet activating factor	PAF
Soluble complement receptor type-1	sCR1
Tissue factor	TF
Tissue plasminogen activator	TPA
Tumor necrosis factor	TNF
United Network for Organ Sharing	UNOS
United States	US
Vaccinia virus complement control protein	VCP

Von willebrand factor	vWF
Xenoreactive natural antibody	XNA

INTRODUCTION

The transplantation of human organs has been a long and continuous challenge triggered by immunological, logistic, and/or economic issues. Our understanding of the immune system in general, the mechanisms involved in allograft rejection and the advancements in drug discovery in particular, have significantly increased since the first human cardiac allotransplant performed by Christian N. Barnard of the University of Cape Town, South Africa in 1967 [1]. Since then, the survival of grafts has been increasingly noticeable. However, the field of transplantation medicine is still subjected to multiple hurdles. For example, according to the United Network for Organ Sharing (UNOS), in the US alone, more than 86,000 patients are currently waiting for a single or multiple organ transplants; however, only 13,223 transplants were performed between January and June of 2004 [2]. As a consequence of this severe organ shortage, many patients continue to rely on conventional therapies. However, the cost of care is generally unaffordable. For example, in the US, the annual expense is up to 35 billion dollars [3]. The use of non-human cells, tissues and organs (xenotransplants) could well be a viable option to surmount the logistic and economic pressures [4,5].

Cellular xenotransplantation has been able to reach into different phases of clinical trials in an attempt to cure various primary cell disorders [6-8]. Different approaches have been tried to treat patients with various neurological ailments such as the amyotrophic lateral sclerosis [6], a degenerative disorder caused by the death of motor neuron cells; Parkinson's disease [9], a progressive neurodegenerative disease characterized by lack of facial mobility and motor coordination, caused by loss of nerve cells; and malignant brain tumors [7], in which therapeutic attempts have been made using genetically engineered murine cells to generate viral vectors expressing the thymidine kinase suicide gene to target cancer cells. It has also been suggested that other progressive neurological disorders such as Alzheimer's and Huntington's disease could be treatable using trophic-factor producing xenogeneic cells [10]. These authors have extensively reviewed the promises of this technology to treat a spectrum of neurological and peripheral ailments [10]. Meanwhile, tissue xenotransplantation has had some preliminary success. The transplantation of cornea, skin, bone marrow and heart valves has been attempted experimentally and/or clinically [11-13]. Due to the relatively minimal immunological and biosafety barriers, cellular and tissue xenotransplantations still have increasing clinical promises. However, the transplantation of solid organs across species has never been an easy challenge either due to the immediate fulminate rejection [14], or the ethical dilemmas associated with accepting animals as organ donors [15].

Non-human primates such as baboons and chimpanzees had been previously used as organ donors clinically [16,17]. Although, by definition, the transplantation of organs from such taxonomically related concordant species does not manifest in hyperacute rejection (HAR) [18], the xenografts were hyperacutely rejected within few days. Moreover, there would be very few organs available from non-human primates, as they are endangered, have single births and a long gestation period. Therefore, researchers have considered non-primates as alternative secondary organ sources.

Among the non-primates, pigs are generally accepted as the most physiologically or morphologically 'compatible' species with the added advantage of short gestation period and large litter size. However, the transplantation of organs from such species taxonomically distant to humans results in a swift rejection process known as hyperacute rejection (HAR), which occurs within minutes to hours depending on the type of organ being transplanted. However, if HAR is diverted successfully, the graft is subjected to subsequent phases of rejection. A brief description of each of the experienced or anticipated types of rejection follows.

TYPES AND MECHANISMS OF XENOREJECTION

Xenotransplantation is defined as the transplantation of cells, tissues and/or organs across species. Organ xenotransplantation is broadly classified into 1) discordant xenotransplantation is the transplantation between phylogenetically distant species such as the guinea pig-to-rat and non-primates-to-humans which manifests in a hyperacute rejection (HAR) [19,20], and 2) concordant xenotransplantation is the transplantation between closely related species such as the mouse-to-rat and non-human primates-to-humans which does not result in hyperacute rejection [21,22].

Hyperacute Rejection (HAR)

HAR is a process used to describe the rapid and aggressive rejection that a vascularized discordant xenograft experiences immediately after an established transplantation. HAR is driven either by the attachment of induced or preformed xenoreactive antibodies (XNAs) to the donor vascular endothelium or by the inability of the natural regulators of complement activation (RCAs) to effectively suppress the spontaneous activation of the alternative complement pathway. For example, in pig-to-human xenograft the preformed antibodies initiate hyperacute rejection amplified by the classical pathway of complement activation [23,24]. This is because humans do not express the functional galactosyl transferase gene; antibodies are naturally preformed against the saccharide gal α 1-3 gal epitopes expressed on the surface of pig endothelial cells [25]. In fact, these xenoantibodies are probably preformed during early childhood against certain α-gal expressing microbes that colonize the gastrointestinal tract [26,27]. This anti-gal α1-3-gal interaction causes activation of the xenograft's endothelial cells and further enhances the activation of the complement system. Figure-1 shows the sequence of events in HAR emphasizing the role of XNAs and complement.

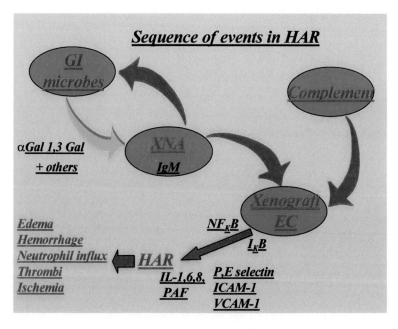

Figure 1. Sequence of events in hyperacute rejection. Illustration of the role of complement and xenoreactive natural antibodies in initiating hyperacute rejection.

In the mouse-to-sensitized rat vascular graft, the induced antibodies drive the classical pathway [28]. However, in a guinea pig-to-rat vascular graft, hyperacute rejection is predominantly due to the failure of the membrane-bound glycoprotein CD59 to effectively bind to C8 and C9 complement components and prevent the spontaneous activation of the rat's heterologous complement [29].

In addition to the XNA and complement, endothelial cell (EC) activation can be a critical component in HAR. Although, in immediate rejections, lasting only a few minutes, endothelial cells do not have the time to become fully activated; in slower rejections lasting several minutes to hours and beyond, the ECs have enough time to be activated and become active contributors in speeding-up the initiated process of rejection.

Upon activation, the cells anticoagulant properties shed and procoagulant properties are gained. Among which, thrombomodulin and heparan sulfate proteoglycans are two important molecules that highly contribute to the physiologic maintenance of the coagulation cascade [30].

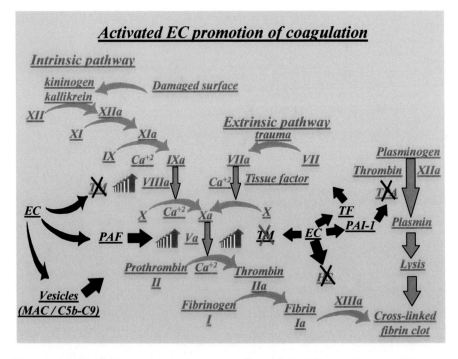

Figure 2. Activated EC promoting coagulation. Illustration of the proteins secreted by activated endothelial cells, which directly affect the coagulation cascade to promote coagulation.

Activated endothelial cells express procoagulant molecules such as the plasminogen activator inhibitor (PAI-1), tissue factor (TF) and platelet-activating factor (PAF). TF and PAF activate factors VII and V respectively, and PAI-1 inhibits tissue plasminogen activator (TPA), a molecule that facilitates the formation of plasmin from its plasminogen precursor (Figure-2). The transcription of thrombomodulin is down regulated and hence factors VIIIa and Va are delayed. A large portion of heparan sulfate is lost from the endothelial boarder [31,32].

Accumulation of C5b-7 complexes on endothelial cells causes the cells to transiently retract from each other [33]. These intracellular gaps formed within the endothelium favour complement activation in the fluid-phase [4,34]. Moreover, the sub-endothelium may be exposed to platelets and Von Willebrand factor (vWF), resulting in platelet activation. The activated platelets then release their prothrombic substances leading to the formation of fibrin clot (Figure-2).

As a result of XNA adherence to the endothelium, hyperactivation of host complement and interrupted coagulation cascade, histopathologic features of HAR follow. They are characterized by interstitial hemorrhage, edema, ischemic necrosis and influx of proinflammatory cells [35-37]. Depending on the model, xenoantibodies attached to vascular endothelium and complement deposition may also be seen with immunological staining.

Accelerated Rejection (AR)

AR is probably named to reflect the accelerated production of antibodies by the recipients of concordant organs such as the mouse-to-naïve-rat and hamster-to-naïve rat. Although there are similarities in the types of cells and mechanisms involved in HAR and AR, there are no preformed antibodies in the latter process. Instead, the antibodies against the xenoendothelial cells are produced 2 to 3 days following transplantation [38].

In AR, CD4+Th2 cells are involved in the activation of B cells through IL-4 secretion, and the stimulated B cells produce IgM and IgG antibodies that are known to have an exclusive role in the rejection process. Here, hyperactivation of the recipient's complement is known to follow the antibody attachment on the graft endothelium. AR mainly differs from HAR in the time of rejection and probably in the predominant class of antibodies involved [38,39], though the predominant antibody in HAR seems to vary depending on the type of organ being transplanted [4, 39,40].

Delayed Xenograft Rejection (DXR)

A delayed phase of rejection (DXR) has been shown to follow hyperacute rejection if HAR is successfully subverted [41-43]. Although it is mainly proinflammatory cells that drive DXR, after averted HAR, residual XNA can still contribute to delayed rejection. DXR, which predominantly involves macrophages and natural killer (NK) cells occurs within 3 to 5 days following xenografting [23,44].

NK cells activate xenoendothelial cells through both antibody-dependent [23,45,46] and complement or antibody-independent (direct) mechanisms [45-47]. In the antibody-dependent cell-mediated cytotoxicity (ADCC), the NK cells recognize the IgG antibody on the surface of 'resting' xenoendothelial cells through their Fc receptor. This exposes the activated ECs to an orchestrated attack by the secreted cytokines from the bound NK cells and macrophages [23,46]. Moreover, Al-Mohanna *et al* [47] have shown the ability of naïve neutrophils to specifically activate xenoendothelial cells independent of XNAs and complement. This *in vitro* study has further addressed the amplified destruction of endothelial cells by naïve NK cells following pre-treatment with neutrophils [47]. Therefore, the role of neutrophils in DXR needs to be elucidated.

The characteristic features of DXR vary from those seen in HAR to the aggregation of inflammatory cells and the secretion of lethal cytokines [41,46,48]. It is believed that DXR is a T-cell-independent phase of rejection [23]. This type of rejection also known as acute vascular rejection (AVR) by some [49, 50] and acute humoral xenograft rejection (AHXR) by others [51] still remains to be a serious challenge in xenotransplantation research.

Acute Cellular Rejection (ACR)

Because discordant xenografts do not usually escape destruction by HAR, AR or DXR, the actual cellular response against the xenograft is not fully understood. However, it is proposed that CD4+T-helper cells are involved in the activation of monocytes, NK cells, and CD8+ Th1 cells all of which have aggravating effect against the xenograft. Moreover, preliminary data in pig-to-human or non-human primate xenotransplantation indicates that human T-cells directly recognize swine xenoantigen on endothelial and dendritic cells. It is speculated that the severity of rejection in ACR is at least equivalent to that of an allograft [49].

Chronic Rejection (CR)

Despite the remarkable progress in haematology and molecular immunobiology, discordant organs have never by-passed the early phases of rejection. Some authors anticipate that chronic xenorejection might be more severe than the corresponding type of rejection in allografts [52]. However, there are palpable concerns about the ability of the xenograft to maintain the expected physiological duty in the recipient milieu. For example, the human liver synthesizes more than 2500 enzymes that are involved in various metabolic activities [53]. Therefore, it is questionable as to whether a pig liver is able to handle the load. It is also worth noting that the normal level of alkaline phosphatase (ALP) in piglets is nine-fold higher than the normal human ALP value [53].

A baboon-to-human liver xenotransplantation has also demonstrated a drop in serum uric acid and cholesterol levels towards the normal baboon profile [54] and nobody is sure about what might happen to the synthesis and kinetics of human red cells following the transplantation of kidneys from pigs, as erythropoietin, a hormone synthesized by the peritubular cells of the kidney and function to stimulate erythroid differentiation and maturation, is not recognized by the human erythropoietin receptors [52]. However, we strongly believe that xenotransplantation research should not be discouraged by such haematological and metabolic incompatibilities and these species-specific variations can only be manipulated experimentally.

APPROACHES TO OVERCOME XENOREJECTION

With the help of various pharmacological, haematological and genetic approaches, the survival of vascularized xenografts has been prolonged from minutes to days. Some of the strategies used to manipulate XNA, complement, and/or endothelial cell activation are emphasized below.

Pharmacologic Drugs

The radical role of immunosuppressive drugs to significantly extend the survival of allografts in the clinic has motivated researchers to evaluate their therapeutic potential in xenotransplantation.

An increasing number of non-primate-to-non-human primate xenotransplants were performed with and without the administration of immunosuppressive drugs since the late 1960s [55]. Pig organs (liver, heart, kidneys and lungs) were transplanted into non-human primates such as the chimpanzee, baboons and monkeys without or accompanied by different immunosuppressive drugs from corticosteroids to cyclosporine and FK506 [51,55,56] in order to mimic the persistence of the xenografts when transplanted into humans. Unfortunately, the survival of the xenografts in the presence or in the absence of these drugs was comparable and short. This is mainly because the allotransplant oriented immunosuppressive drugs target cellular response and lack the ability to prevent the xenograft from XNA driven HAR [55]. Not only are immunosuppressive drugs ineffective when used alone, but also unlike in allografts the drug regimen used in xenograft recipients is heavy [17] and can easily expose the patient to microbial infections by attenuating any residual defense mechanism against pathogenic microorganisms. Therefore, it was crucial to combine this approach with other strategies that can deplete the recipient's XNA or genetic engineering techniques to generate modified animals.

Haematological Approaches

To date, several haematological strategies have been applied to overcome HAR in different animal models [57-59]. Xenoantibodies have been significantly depleted by exchanging the recipient's plasma with isotonic solutions [57,60]. Yet, this approach also exhausts many vital proteins, and therefore column-based techniques that specifically bind the reactive xenoantibodies have been used subsequently [60,61].

Transfusion techniques have also been used either to remove XNAs from the recipient's circulation [62] or to induce graft tolerance by reducing the B cell mediated response [63]. The xenoperfusion approach has been suggested to be useful in xenotransplantation experiments [64] and Tuso *et al* [62] obtained high titre reduction in the IgG and IgM xenoantibodies after perfusing human blood through a pig liver or kidney. Moreover, Bersztel *et al* [63] have achieved partial graft tolerance after transfusing a rat with lymphocyte enriched mice blood prior to cardiac xenotransplantation.

Bone marrow transplantation or infusion of haematopoietic cells has also been attempted to induce some sort of tolerance [58,59]. It was then suggested that the phagocytic action of macrophages was responsible for the eventual loss of the infused/transplanted cells [58] and the attempt to deplete circulating macrophages was not successful. However, Takayama et al [52] have reported the successful abrogation of macrophages using the intravenous administration of Clodronate. However, none of these haematological approaches is adequate to abolish HAR indefinitely.

Molecular Approaches

The application of molecular techniques in transplantation research can be considered as a tangible step towards the reality of clinical xenotransplantation. It has been possible to engineer animals lacking the gal α 1-3 gal gene [25] and to generate transgenic animals expressing different membrane-bound human complement regulatory proteins [18,65-68]. They all have been shown to significantly prolong survival of xenografts [69-71]. For example, it has been inferred that expression of the complement regulator decay-accelerating factor (DAF) (CD55) makes a discordant organ function as an organ from a concordant species [72]. Moreover, the expression of membrane cofacotor protein (MCP) (CD46) and the membrane inhibitor of reactive lysis (CD59) in mice and pigs have confirmed their therapeutic potential [18,65].

Interestingly, the expression of a glycosyltransferase, $\alpha 1,2$ fucosyltransferase, which competes with the gal α 1-3 galactosyltransferase to transfer galactose sugar to the common substrate N-acetyllactosamine during the synthesis of certain carbohydrates [24,73], has been shown to greatly reduce the level of gal α 1-3 gal expression both in vitro [73] and in transgenic mice [24,73] and pigs [73]. We believe that this strategy is as promising as the generation of gal α 1-3 knock out pigs. Moreover, the transfection of human cells with a shuttle vector carrying the human α-galactosidase gene has also been shown to attenuate gal α 1-3 gal expression and was suggested to have promises in gene therapy based xenotransplantation research [74].

Although the various molecular approaches reviewed above have continued to show their potential role experimentally, some of them may not be sufficient to be applied independently for clinical use. For example, some studies speculated that the degree to which DAF regulates hyperactivation of the complement system is not adequate for its use in the clinic [75-77]. Other studies have reported that the alternative pathway can still remain unregulated despite the presence of membrane-anchored DAF and MCP [78] and because these membrane bound proteins are expressed in clusters, they fail to control the activation of complement in the fluid-phase, which may account for complement mediated xenorejection [4,34]. To effectively regulate complement and abolish hyperacute rejection, other therapeutic approaches have been elucidated.

Therapeutic Approaches

Different soluble complement control peptides and proteins have been used to regulate hyperactivation of the classical, alternative or both pathways of complement activation. Among which the peptide Compstatin [4] and proteins such as the soluble complement receptor type 1 (sCR1) [79-81], cobra venom factor (CVF) [38,82] and the vaccinia virus complement control protein (VCP) [83-85] can be mentioned.

Compstatin, a 27 amino acid peptide isolated from a phage library, can block the complement component C3 [86] and hence effectively regulates the alternative pathway. The use of this peptide is suggested to be feasible in models where the alternative pathway predominates HAR [4] as in the guinea pig-to-rat. However, it lacks the ability to significantly inhibit the classical pathway, platelet and leukocyte activations [4], and therefore has limited potential.

The soluble complement receptor type 1 (sCR1) is highly effective to abrogate complement activation by inhibiting both the classical and the alternative pathways [79,81]. Pruitt *et al* [81] have proven the tremendous ability of sCR1 to prolong pig-to-cynomolgus monkey cardiac xenotransplant for up to 4days compared to the rapid and aggressive rejection (within 1hour) in the control groups. However, because of its bulky size (120kDa) it is likely to induce a complex immune response when administered continuously.

Cobra venom factor (CVF) is also a glycoprotein with an extreme potency to abolish complement activation [38,82]. Its potential has been evaluated in guinea pig-to-rat [82] and pig-to-baboon [38] models. Surprisingly, in the guinea pig-to-rat cardiac xenotransplants, CVF treated hearts survived above 295-fold longer than the untreated control groups, which hyperacutely rejected in about 20 minutes [82]. However, CVF mainly inhibits the alternative pathway of complement activation and therefore may not be as effective in models where XNA or the classical pathway drive the hyperacute rejection. Moreover, the presence of diffused hemorrhage and aggregated inflammatory cells [82] suggests its limitations to control subsequent phases of xenorejection.

The other soluble complement modulatory protein with exciting potential in preventing xenorejection is the vaccinia virus complement control protein (VCP). This vaccinia virus encoded protein [83] was the first microbial protein to have a nominated role in regulation of the complement system and viral escape from example, in Japan among the followers of Shinto, death is considered as a misfortune and the dead body is treated as infectious and unclean and therefore any of its parts should not be maimed [97]. Due to this and other causes of limited allo-organ logistics, the medical community has been forced to consider xenotransplantation as an alternative life support.

Despite the significant scientific progress in xenotransplantation as described earlier in the chapter, and some encouraging support from the religious communities [98], the field is still surrounded with unresolved ethical issues. For example, there are concerns of social stigma to the xenograft recipients due to the 'unusual' transplant [96]. Moreover, the animal rights groups do not support the 'restoring life at the expense of another life' approach, although this argument can be strongly challenged by pointing to the extensive consumption of animals for food purposes [99]. Furthermore, some people are generally against the incorporation of xenografts into humans due to the fear of exposing the public to a number of animal derived microbial infections [100-102], commonly called xenozoonosis. For example, Nipah virus, a member of the paramyxovirinae family, has been shown to establish encephalitis in humans in Singapore and Malaysia [103]. Since 1998, this pig-derived neurovirulent virus has widely infected abattoirs and other people closely working with pigs. Therefore, there is a need to closely monitor the paramyxoviruses and other families of swine viruses [104] to prevent the potential risk of xenozoonosis through xenotransplantation.

CONCLUDING REMARKS

We believe that if we formulate carefully the aforementioned dynamic approaches, the hyperacute and the delayed phases of xenorejection may soon be surmounted and the anticipated chronic xenorejection may be prevented with the help of immunosuppressive drugs and other therapeutic reagents. We remain optimistic to see the medical, ethical and moral dimensions to work synergistically towards the promising clinical benefits.

ACKNOWLEDGEMENTS

The Poliomyelitis Research Foundation (PRF), and the University of Cape Town support YTG. Jewish Hospital Research Foundation is gratefully acknowledged for the research support on xenotransplantation in GJK's lab. GJK is currently a senior International Wellcome Trust fellow in Biomedical Sciences in South Africa.

REFERENCES

[1] Barnard C.N. (1967) A human cardiac transplant: An interim report of a successful operation performed at Groote Schuur Hospital, Cape Town. *S. A. Medical Journal* 41, 1271-1274.
[2] United Network for Organ Sharing Homepage: *www.unos.org*
[3] Michler R.E. (1996) Xenotransplantation: Risks, Clinical potential, and Future Prospects. *Emerging Infectious Diseases* 2(1), 64-70.
[4] Mollnes T.E. and Fiane A.E. (1999) Xenotransplantation: how to overcome the complement obstacle? *Molecular immunology* 36, 269-276.
[5] Daar A.S. (2003) Xenotransplantation: Recent Scientific Developments and Continuing Ethical Discourse. *Transplantation Proceedings* 35, 2821-2822.
[6] Aebischer P., Schluep M., Déglon N., Joseph J-M., Hirt L., Heyd B., Godard M., Hammang J., Zurn A.D., Kato A.C., Regli F. and Baetge E. (1996) Intrathecal delivery of CNTF using encapsulated genetically modified xenogeneic cells in amyotrophic lateral sclerosis patients. *Nature Medicine* 2(6), 696-699.
[7] Ram Z. *et al* (1997) Therapy of malignant brain tumors by intratumoral implantation of retroviral vector-producing cells. *Nature Medicine* 3(12), 1354-1361.
[8] Aebischer P., Hottinger A.F. and Déglon N. (1999) Cellular xenotransplantation. *Nature Medicine* 5(8), 852.
[9] Deacon T., Schumacher J., Dinsmore J., Thomas C., Palmer P., Kott S., Edge A., Penney D., Kassissieh S., Dempsey P. and Isacson O. (1997) Histological evidence of fetal pig neural cell survival after transplantation into a patient with Parkinson's disease. *Nature Medicine* 3(3), 350-353.
[10] Lanza R.P. and Cooper D.K.C. (1998) Xenotransplantation of cells and tissues: application to a range of diseases, from diabetes to Alzheimer's. *Molecular medicine today* 4, 39-45.
[11] Lehrman S. (1995) AIDS patient given baboon bone marrow. *Nature* 378, 756.
[12] Ohno K, Nelson L.R., Mitooka K. and Bourne W.M. (2002) Transplantation of cryopreserved human corneas in a xenograft model. *Cryobiology* 44(2), 142-149.

[13] Nickoloff B.J. and Nestle F.O. (2004) Recent insights into the immunopathogenesis of psoriasis provide new therapeutic opportunities. *Journal of Clinical Investigations* 113(12), 1664-1675.
[14] Platt J.L., *et al* (1990) Transplantation of discordant xenografts: a review of progress. *Immunology Today* 11(12), 450-457.
[15] The Right Reverend Lord John Habgood, Spagnolo A.G., Sgreccia E. and Daar A.S. (1997) Religious Views on Organ and Tissue Donation. In Jeremy R. Chapman, Mark Deierhoi and Celia Wight (Eds.) *Organ and Tissue Donation for Transplantation* (pp. 23-33). Arnold, a member of the Hodder Headline Group.
[16] Reemtsma K. (1991) Xenotransplantation – A Brief History of Clinical Experiences: 1900-1965: Early Attempts at Renal Xenografting. In D.K.C. Cooper, E. Kemp, K. Reemtsma, and D.J.G. White (Eds.) *Xenotransplantation: The Transplantation of Organs and Tissues Between Species* (pp. 9-22). Springer-Verlag Berlin Heidelberg.
[17] Barnard C.N., Wolpowitz A. and Losman J.G. (1977) Heterotopic Cardiac Transplantation with a Xenograft for Assistance of the Left Heart in Cardiogenic Shock after Cardiopulmonary Bypass. *S. Afr. Med. J.* 52, 1035-1038.
[18] Bach F.H. and Auchincloss H. (1995) *Transplantation immunology*. New York: Wiley-Liss. Xi, 409.
[19] Cooper D.K.C. (1991) Xenotransplantation: *The Transplantation of Organs and Tissues Between Species*. Berlin; New York: Springer-Verlag. Xxix, 583.
[20] Luo Y., *et al* (1998) Comparative histopathology of hepatic allografts and xenografts in the nonhuman primate. *Xenotransplantation* 5(3), 197-206.
[21] Galili U., *et al* (1988) Man, apes, and Old World monkeys differ from other mammals in the expression of alpha-galactosyl epitopes on nucleated cells. *J Biol Chem* 263(33), 17755-17762.
[22] Galili U. (1998) Anti-Gal antibody prevents Xenotransplantation. *Science and Medicine* 5, 28-37.
[23] Takahashi M., Nakajima S., Miyajima K., Ogata K., Suzuki A., Konaka C. and Kato H. (2002) Role of Xenoreactive Natural Antibodies in Pig-to-Human Lung Xenotransplantation. *Transplantation Proceedings* 34, 2739-2744.
[24] Shinkel T.A., Chen C-G., Salvaris E., Henion T.R., Barlow H., Galili U., Pearse M.J. and D'apice A.J.F. (1997) Changes in cell surface glycosylation in α1,3-galactosyltransferase knockout and α1,2-fucosyltransferase transgenic mice. *Transplantation* 64(2), 197-204.

[25] Lai L., Kolber-Simonds D., Park K-W., Cheong H-T., Greenstein J.L., Im G-S., Samuel M., Bonk A., Rieke A., Day B.N., Murphy C.N., Carter D.B., Hawley R.J. and Prather R.S. (2002) Production of α-1,3-Galactosyltransferase Knockout Pigs by Nuclear Transfer Cloning. *Science* 295, 1089-1092.

[26] Galili U., *et al* (1988) Interaction between human natural anti-alpha-galactosyl immunoglobulin G and bacteria of the human flora. *Infect Immun.* 56(7), 1730-1737.

[27] Bracy J.L., Sachs D.H. and Iacomini J. (1998) Inhibition of Xenoreactive Natural Antibody Production by Retroviral Gene Therapy. *Science* 281, 1845-1847.

[28] Anderson J.B., Smith S.A., van Wijk R., Chien S. and Kotwal G.J. (2002) Vaccinia Virus Complement Control Protein Ameliorates Hyperacute Xenorejection by Inhibiting Xenoantibody Binding. *Transplantation Proceedings* 34, 3277-3281.

[29] Anderson J.B., Smith S.A., van Wijk R., Chen S. and Kotwal G.J. (2003) Vaccinia virus complement control protein inhibits hyperacute xenorejection in a guinea pig-to-rat heterotopic cervical cardiac xenograft model by blocking both xenoantibody binding and complement pathway activation. *Transplant immunology* 11(2), 129-135.

[30] Platt J.L. and Bach F.H. (1991) Mechanisms of Tissue Injury in Hyperacute Xenograft Rejection: A Model of Hyperacute Xenograft Rejection. In D.K.C. Cooper, E. Kemp, K. Reemtsma, and D.J.G. White (Eds.) *Xenotransplantation: The Transplantation of Organs and Tissues Between Species* (pp. 69-79). Springer-Verlag Berlin Heidelberg.

[31] Platt J.L., *et al* (1990) Release of heparan sulfate from endothelial cells. Implications for pathogenesis of hyperacute rejection. *J Exp Med* 171(4), 1363-1368.

[32] Ihrcke N.S. and Platt J.L. (1996) Shedding of heparan sulfate proteoglycan by stimulated endothelial cells: evidence for proteolysis of cell-surface molecules. *J Cell Physiol* 168(3), 625-637.

[33] Saadi S. and Platt J.L. (1995) Transient perturbation of endothelial integrity induced by natural antibodies and complement. *J Exp Med* 181(1), 21-31.

[34] Fiane A.E., Videm V., Lambris J.D., Geiran O.R., Svennevig J.L. and Mollnes T.E. (2000) Modulation of Fluid-Phase Complement Activation Inhibits Hyperacute Rejection in a Porcine-to-Human Xenograft Model. *Transplantation Proceedings* 32, 899-900.

[35] Haisch C.E., *et al* (1990) The vascular endothelial cell is central to xenogeneic immune reactivity. *Surgery* 108(2), 306-311.

[36] Rose A.G., et al (1991) Histopathology of hyperacute rejection of the heart: experimental and clinical observations in allografts and xenografts. *J Heart Lung Transplant* 10(2), 223-234.
[37] Rose A.G. and Cooper D.K.C. (1996) A histopathologic grading system of hyperacute (humoral, antibody-mediated) cardiac xenograft and allograft rejection. *J Heart Lung Transplant* 15(8), 804-817.
[38] Takayama J., Koyamada N., Abe T., Hatsugai K., Usuda M., Ohkohchi N. and Satomi S. (2000) Macrophage Depletion Prevents Accelerated Rejection and Results in Long-Term Survival in Hamster to Rat Cardiac Xenotransplantation. *Transplantation Proceedings* 32, 1016.
[39] Schraa E.O. et al (1999) IgG, but not IgM, mediates hyperacute rejection in hepatic xenografting. *Xenotransplantation* 6, 110-116.
[40] Dehoux J.P., Hori S., Talpe S., Bazin H., Latinne D., Soares M.P. and Gianello P. (2000) Specific depletion of preformed IgM natural antibodies by administration of anti-mu monoclonal antibody suppresses hyperacute rejection of pig to baboon renal xenografts. *Transplantation* 70, 935-946.
[41] Kobayashi T., et al (1997) Delayed xenograft rejection of pig-to-baboon cardiac transplants after cobra venom factor therapy. *Transplantation* 64(9), 1255-1261.
[42] Leventhal J.R., Dalmasso A.P., Cromwell J.W., Platt J.L., Manivel C.J., Bolman III R.M. and Matas A.J. (1993) Prolongation of cardiac xenograft survival by depletion of complement. *Transplantation* 55(4), 857-866.
[43] Bach F.H., Robson S.C., Winker H., Ferran C., Stuhlmeier K.M., Wrighton C.J. and Hancock W.W. (1995) Barriers to xenotransplantation. *Nature Medicine* 1(9), 869-873.
[44] Bach F.H., Robson S.C., Ferran C., Millan M., Anrather J., Kopp C., Lesnikoski B., Goodman D.J., Hancock W.W., Wrighton C. and Winker H. (1995) Xenotransplantation: Endothelial cell Activation and Beyond. *Transplantation Proceedings* 27(1), 77-79.
[45] Luca I., Samaja M., Motterlini R., Mangili F., Bender J.R. and Pardi R. (1992) Early recognition of a discordant xenogeneic organ by human circulating lymphocytes. The *Journal of Immunology* 149(4), 1416-1423.
[46] Goodman D.J., Albertini M., Willson A., Millan M.T. and Bach F.H. (1996) Direct activation of porcine endothelial cells by human natural killer cells. *Transplantation* 61(5), 763-771.
[47] Al-mohanna F. et al (1997) Activation of Naïve Xenogeneic but Not Allogeneic Endothelial Cells by Human Naïve Neutrophils. A Potential Occult Barrier to Xenotransplantation. *American Journal of Pathology* 151(1), 111-120.

[48] Kozlowski T., et al (1999) Porcine kidney and heart transplantation in baboons undergoing a tolerance induction regimen and antibody adsorption. *Transplantation* 67(1), 18-30.
[49] Cooper D.K.C. and Platt J.L. (1997) Xenotransplantation: *The Transplantation of Organs and Tissues between species*. 2nd ed. Berlin; New York: Springer. Xxxiv, 854.
[50] Platt J.L. (1998) Current Status of Xenotransplantation: Research and Technology. *Transplantation Proceedings* 30, 1630-1633.
[51] Holmes B.J. et al (2001) Antibody Responses in early Graft Rejection in Pig-to-Primate Renal Xenotransplantation. *Transplantation Proceedings* 33, 717-718.
[52] Seow J. (2003) Clinical Xenotransplantation. *The Lancet* 362, 1421-1422.
[53] Munitiz V. et al (2002) Analytical Profile Comparison Between Pig and Baboon in an Orthotopic Liver Xenotransplantation Model. *Transplantation Proceedings* 34, 323-324.
[54] Starzl T.E. et al (1993) Baboon-to-human liver transplantation. *The Lancet* 341(8837), 65-71.
[55] Lambrigts D., Sachs D.H. and Cooper D.K.C. (1998) Discordant Organ Xenotransplantation in Primates. *Transplantation* 66(5), 547-561.
[56] Makowka L. et al (1995) The use of a pig liver xenograft for temporary support of a patient with fulminate hepatic failure. *Transplantation* 59(12), 1654-1659.
[57] Reding R., Davies S. ff. H., White D.J.G., Wright L.J., Marbaix E., Alexander G.P.J., Squifflet J.P. and Calne R.Y. (1989) Effect of Plasma Exchange on Guinea pig-to-Rat Heart Xenografts. *Transplantation Proceedings* 21(1), 534-536.
[58] Cooper D.K.C., Gollackner B., Knosalla C. and Teranishi K. (2002) Xenotransplantation – how far have we come? *Transplant immunology* 9, 251-256.
[59] Asano M., Gundry S.R., Izutani H., Cannarella S.N., Fagoaga O. and Bailey L.L. (2003) Baboons undergoing orthotopic concordant cardiac xenotransplantation surviving more than 300 days: Effect of immunosuppressive regimen. *The Journal of Thoracic and Cardiovascular Surgery* 125, 60-70.
[60] Bach F.H., Platt J.L. and Cooper D.K.C. (1991) Accommodation – The Role of Natural Antibody and Complement in Discordant xenograft Rejection: Methods of Removal of Preformed Natural Antibodies. In D.K.C. Cooper, E. Kemp, K. Reemtsma, and D.J.G. White (Eds.)

Xenotransplantation: The Transplantation of Organs and Tissues Between Species (pp. 81-99). Springer-Verlag Berlin Heidelberg.
[61] Sachs D., Sykes M., Greenstein J.L. and Cosimi A.B. (1995) Tolerance and xenograft Survival. *Nature Medicine* 1(9), 969.
[62] Tuso P.J., Carmer D.V., Yasunaga C., Cosenza C.A., Wu G.D. and Makowka L. (1993) Removal of natural human xenoantibodies to pig vascular endothelium by perfusion of blood through pig kidneys and livers. *Transplantation* 55(6), 1375-1378.
[63] Bersztel A., Johnsson C., Björkland A. and Tufveson G. (2003) Pretransplant Xenogeneic Blood Transfusions Reduce the Humoral Response in a Mouse-to-Rat Heart Transplantation Model. *Scandinavian Journal of Immunology* 57, 246-253.
[64] Otte K.E., Steinbrüchel D., Kain H. and Kemp E. (1992) Xenoperfusion Experiments Are Suitable for Xenotransplantation Research. *Transplantation Proceedings* 24(2), 449-450.
[65] McCurry K.R., *et al* (1995) Human complement regulatory proteins protect swine-to-primate cardiac xenografts from humoral injury. *Nature Medicine* 1(5), 423-427.
[66] Langford G.A., Yannoutsos N., Cozzi E., Lancaster R., Elsome K., Chen P., Richards A. and White D.J.G. (1994) Production of Pigs Transgenic for Human Decay Accelerating Factor. *Transplantation Proceedings* 26(3), 1400-1401.
[67] Cozzi E. and White D.J.G. (1995) The generation of transgenic pigs as potential organ donors for humans. *Nature Medicine* 1(9), 964-966.
[68] Rosengard A.M., Carry N.R.B., Langford G.A., Tucker A.W., Walluork J. and White D.J.G. (1995) Tissue expression of human complement inhibitor, decay-accelerating factor, in transgenic pigs. *Transplantation* 59(9), 1325-1333.
[69] Goddard M.J., Dunning J.J., Horsley J., Atkinson C., Pino-Chavez G. and Wallwork J. (2002) Histopathology of Cardiac Xenograft Rejection in the Pig-to-Baboon Model. *J Heart Lung Transplant* 21, 474-484.
[70] Brandl U., *et al* (2003) 25 days of baboons after orthotopic xenotransplantation of hDAF transgenic pig hearts using a moderate immunosuppression (Abstract) *The Journal of Heart and Lung Transplantation* 22(1S), S100-S101.
[71] Ramirez P., *et al* (2002) Transgenic Pig-to-Baboon Liver Xenotransplantation: Clinical, Biochemical, and Immunologic Pattern of Delayed Acute Vascular Rejection. *Transplantation Proceedings* 34, 319-320.

[72] Schmoeckel M., *et al* (1997) Transgenic human decay accelerating factor makes normal pigs function as a concordant species. *J Heart Lung Transplant* 16(7), 758-764.
[73] Sharma A. *et al* (1996) Reduction in the level of Gal (α1,3)Gal in transgenic mice and pigs by the expression of an α(1,2)fucosyltransferase. *Proc. Natl. Acad. Sci. USA* 93, 7190-7195.
[74] Jing-Lian Y., Lu-Yang Y., Li-Hua Z. and Li-He G. (2003) Expression of human α-galactosidase leads to reduction of major xenoepitope Galα(1,3)Gal in NIH 3T3 cell. *Acta Pharmacologica Sinica* 24(10), 985-990.
[75] van Denderen B.J., *et al* (1997) Combination of decay-accelerating factor expression and alpha 1,3-galactosyltransferase knockout affords added protection from human complement-mediated injury. *Transplantation* 64(6), 882-888.
[76] Koike C., *et al* (1996) Establishment of a human DAF/HRF20 double transgenic mouse line is not sufficient to suppress hyperacute rejection. *Surg Today* 26(12), 993-998.
[77] Makrides S.C. (1998) Therapeutic inhibition of the complement system. *Pharmacol Rev* 50(1), 59-87.
[78] Kotwal G.J. (1997) Microogranisms and their interaction with the immune system. *Journal of Leukocyte Biology* 62(4), 415-429.
[79] Ryan U.S. (1995) Complement inhibitory therapeutics and Xenotransplantation. *Nature Medicine* 1(9), 967-968.
[80] Xia W., *et al* (1992) Prolongation of guinea pig cardiac xenograft survival in rats by soluble human complement receptor type 1. *Transplantation Proceedings* 24, 479-480.
[81] Pruitt S.K., *et al* (1994) The effect of soluble complement receptor type 1 on hyperacute rejection of porcine xenografts. *Transplantation* 57, 363-370.
[82] Sun Q-Y., Chen G., Guo H., Chen S., Wang W-Y. and Xiong Y-L (2003) Prolonged cardiac xenograft survival in guinea pig-to-rat model by a highly active cobra venom factor. *Toxicon* 42(3), 257-262.
[83] Kotwal G.J. and Moss B. (1988) Vaccinia virus encodes a secretory polypeptide structurally related to complement control proteins. *Nature* 335, 176-178.
[84] Kotwal G.J., Isaacs S.N., Mckenzie R., Frank M.M. and Moss B. (1990) Inhibition of the Complement Cascade by the Major Secretory Protein of Vaccinia Virus. *Science* 250, 827-829.
[85] Isaacs S.N., Kotwal G.J. and Moss B. (1992) Vaccinia virus complement-control protein prevents antibody-dependent complement-enhanced

neutralization of infectivity and contributes to virulence. *Proc Nat Acad Sci USA* 89, 628-632.
[86] Sahu A., Kay B.K. and Lambris J.D. (1996) Inhibition of Human Complement by a C3-Binding Peptide Isolated from a Phage-Displayed Random Peptide Library. *Journal of Immunology* 157, 884-891.
[87] Daly J. and Kotwal G.J. (1998) Pro-inflammatory complement activation by the A beta peptide of Alzheimer's disease is biologically significant and can be blocked by vaccinia virus complement control protein. *Neurobiology of Aging* 19, 619-627.
[88] Hicks R.A., Keeling K.L., Yang M-Y., Smith S.A., Simons A.M. and Kotwal G.J. (2002) Vaccinia Virus Complement Control Protein Enhances Functional Recovery after Traumatic Brain Injury. *Journal of Neurotrauma* 19(6), 705-714.
[89] Scott M.J., Burch P.T., Jha P., Peyton J.C., Kotwal G.J. and Cheadle W.G. (2003) Vaccinia Virus Complement Control Protein Increases Early Bacterial Clearance during Experimental Peritonitis. *Surgical Infections* 4(4), 317-326.
[90] Kotwal G.J. (2000) Poxviral mimicry of complement and chemokine system components: what's the end game? *Immunology Today* 21(5), 242-248.
[91] Al-Mohanna F., Parhar R. and Kotwal G.J. (2001) Vaccinia virus complement control protein is capable of protecting xenoendothelial cells from antibody binding and killing by human complement and cytotoxic cells. *Transplantation* 71(6), 796-801.
[92] Kahn D., Smith S.A. and Kotwal G.J. (2003) Dose-Dependent Inhibition of Complement in Baboons by Vaccinia Virus Complement Control Protein: Implications in Xenotransplantation. *Transplantation Proceedings* 35, 1606-1608.
[93] Jha P., Smith S.A., Justus D.E. and Kotwal G.J. (2003) Prolonged Retention of Vaccinia Virus Complement Control Protein Following IP Injection: Implications in Blocking Xenorejection. *Transplantation Proceedings* 35, 3160-3162.
[94] Reynolds D.N. *et al* (2000) Heparin binding activity of vaccinia virus complement control protein confers additional properties of uptake by mast cells and attachment to endothelial cells; In *Advances in animal virology* (ed.) S Jameel (Villarreal: Science Publishers) pp 337-342.
[95] White D.J.G. (1997) Xenotransplantation – a solution to the donor organ shortage. In Jeremy R. Chapman, Mark Deierhoi and Celia Wight (Eds.) *Organ and Tissue Donation for Transplantation* (pp. 446-457). Arnold, a member of the Hodder Headline Group.

[96] Wright R.A. (1991) An Ethical Framework for Considering The Development of Xenotransplantation in Man. In D.K.C. Cooper, E. Kemp, K. Reemtsma, and D.J.G. White (Eds.) *Xenotransplantation: The Transplantation of Organs and Tissues Between Species* (pp. 511-527). Springer-Verlag Berlin Heidelberg.
[97] Namihira E. (1990) Shinto Concept Concerning the Dead Human Body. *Transplantation Proceedings* 22(3), 940-941.
[98] McKay D. (2001) The Vatican and Xenotransplantation. *TRENDS in Biotechnology* (News & Comment) 19(12), 489.
[99] Prather R.S., Hawley R.J., Carter D.B., Lai L. and Greenstein J.L. (2003) Transgenic swine for biomedicine and agriculture. *Theriogenology* 59, 115-123.
[100] Kalter S.S. (1991) The Nonhuman Primate as Potential Organ Donor for Man: Virological *Xenotransplantation: The Transplantation of Organs and Tissues Between Species* (pp. 457-479). Springer-Verlag Berlin Heidelberg.
[101] Bach F.H., *et al* (1998) Uncertainty in xenotransplantation: Individual benefit versus collective risk. *Nature Medicine* 4(2), 141-144.
[102] Allan J.S. (1996) Xenotransplantation at a crossroads: Prevention versus progress. *Nature Medicine* 2(1), 18-21.
[103] Wong K.T., Shieh W.J., Zaki S.R. and Tan C.T. (2002) Nipah virus infection, an emerging paramyxoviral zoonosis. *Springler Seminars in Immunopathology* 24, 215-228.
[104] Paul P.S., Halbur P., Janke B., Joo H., Nawagitgul P., Singh J. and Sorden S. (2003) Exogenous porcine viruses. *Current topics in microbiology & immunology* 278, 125-183.

In: Contemporary Ethical Issues
Editor: Albert G. Parkis, pp. 87-96

ISBN 1-59454-536-7
© 2006 Nova Science Publishers, Inc.

Chapter 5

NEW FINDINGS ON EARLY EMBRYO RESEARCH AND THEIR ETHICAL RELEVANCE

Miguel Ruiz-Canela[*]
University of Navarra Irunlarrea,
Pamplona, Spain

ABSTRACT

Recent research on the human embryo is revealing us surprising insights on the biology of the earliest stages of our own life. The description of the changes -- molecular, genetic or structural -- taking place from fertilization to blastocyst formation makes a story that fills us with wonder and amazement.

The novelty is twofold. First, it reveals us the untold variety, exactness and fine integration of the mechanisms and processes involved, and replaces the old crude sketch of spherical zygotes and identical blastomeres with a lively and elaborate depiction of dissimilar cells endowed with axis and poles, faces and asymmetries. Second, the new biological findings bring with them some factual data that add fresh impetus and deeper understanding to the enduring controversy on the ethical status of the early human embryo.

---

[*] Miguel Ruiz-Canela Biomedical Humanities Faculty of Medicine University of Navarra Irunlarrea, 1 E-31008 Pamplona. SPAIN Fax: (34)-48-425649 Tel: (34)-48-425600 e-mail: mcanela@unav.es

INTRODUCTION

The status of the human embryo as a research subject has been the object of current controversy in the United States (Weiss 2002; Check 2002). In fact, the growing expansion of this field acts as a constant stimulus to ethical reflection: the use of embryo stem cells and cloning are being the subject of an intense argumentation in which ethical, legal, political and socio-economic aspects intersect (Lauritzen 2001). Moreover, recent research with human embryos has been very controversial even by experts in Human Reproduction and Embryology (Hutchinson 2003). Apart from legal control, research possibilities in this field seem to be limited only by scientists' imagination, as we can see in a meeting of American and Canadian biologists on whether to recommend stem cell experiments that would mean creating a human-mouse hybrid (Wade 2002).

In this regard, the debate on some extreme research raises a new issue, that is, the status that should be conferred to the embryo resulting from some types of research like somatic cell nuclear transfer (SCNT), parthenogenesis or even the creation of animal-human hybrids. In many cases these embryos are non-viable and it could even be thought that they are not real embryos at all (Bruce 2002; McHugh 2004). This conclusion would lead to radically different ethical consequences and it would then require a less strict control over it.

This paper aims to contextualise some new advances in the evolving field of molecular embryology. Significant findings in embryology during recent years have added important data that can be very useful to elucidate the characteristics that must be present in an early human embryo.

GENOME AND EPIGENETICS

Once the spermatozoon penetrates the zona pellucida surrounding the oocyte, several rapid processes occur until the female and male pronuclei fuse to produce the new, diploid nucleus of the zygote (Evans and Florman 2002). So, the single cell embryo inherits the genetic information from both parents, constitutes its own genome, where the information required to start and guide the whole development process is contained.

Genetic information for the development of the embryo should not be understood in a fixed and excluding way; we must consider the increasingly important process of epigenetic modification (Pennisi 2001; Van Speybroeck 2002). Epigenetics refers to all processes relating to the expression of genes that are determined by events beyond the level of genetic information (Oligny 2001).

It tells us that gene expression depends on surrounding contexts: intracellular, intercellular, organismic, and environmental. Epigenetics explains why discrete cells appear in the development process of the zygote although every cell has the same genetic information. As a consequence, the hereditary and developmental processes involve more than the genes themselves, and require an ongoing interplay between embryo and environment.

Patterns of gene expression, not genes themselves, define each cell type (Pennisi 2001). There are several mechanisms that turn genes "on" and "off" where intracellular Ca^{2+} plays an important role, and chromatin structural changes, as well as DNA methylation, assemblage of histone proteins into nucleosomes, and remodeling of chromosome associated proteins are also involved (Nakao 2001). These changes imply the modification and even the amplification of the information enclosed in the DNA.

Accordingly, there are two levels of information, genomic imprinting and epigenetic regulation, and both of them contribute to the development of different phenotypes (embryonic, fetal, neonatal, etc.) in a continuous sequence. Differentiation is regulated by cascades of gene expression and in this process there are memory mechanisms involving DNA modifications such as DNA methylation and chromatin remodeling which inhibit or activate the expression of genes (Nakao 2001). These changes are transmitted to daughter cells and can be considered as a cell-watch where a higher number of cytosines methylated implies an older organism (Oligny 2001). Differential growth through epigenetic influences allows specific gene expression needed to obtain different tissues, a process of self-construction which curiously begins with the zygote formation. At the same time, several random changes in cellular and intercellular regulatory network, such as fluctuations in the concentration and location of messengers, also contribute to the determination of phenotype in each stage. Phenotypic variation is the result of both genetic activation and environmental influences.

However, a group of specialized cells is not necessarily a living organism. For example, a teratoma, which may possess several cell types, lacks the capacity to form a new entity. Apart from epigenetically regulated gene expression, another condition is required: differential development of cells has to be co-ordinated to allow these cells to constitute an individual. Accordingly, a pattern formation is needed, which allows a sequence of ordered messages co-ordinated in time and space. Processes of embryonic cell growth consist of extensive and accurately orchestrated cell proliferation, differentiation, and death (Oligny 2001). The result of the developmental program is the series of complex processes of spatially and temporally ordered cell differentiation, directed cell movement, and shaping of structures composed of several cell types. At the same time, all embryonic cells share the same genetic information and this preserves the identity of the living organism.

PHENOTYPE-ZYGOTE: A LIVING ORGANISM

The main issue at stake is identifying the moment in which the aforementioned characteristics that configure a living organism appear in an embryo. For the past 30 years, it was usually assumed that axes determination in mammals were due to some type of external signal since no clear indications had been found for a bilateral symmetry in zygotes and early blastocysts. Although several studies had showed that mammalian zygotes possess cytoplasmic determinants, which are segregated unequally to the various blastomeres during cleavage, these studies failed to emphasize the fact that axis determination is usually connected with determination of cell fate (Denker 2004). In fact, until 1989, it was assumed that gastrulation (the process by which germ layers in mammalian embryo are formed) was the most significant moment in embryo development, based on the appearance of polarity.

In this regard, evolving knowledge about mechanisms underlying the patterning information in early mammalian embryos can be helpful (Johnson 2001; Pearson 2002). Recent studies show how well-known principles of axis determination described in amphibia are in many respects also at work in mammals (Denker 2004). Developmental biologists have found that after the first division of a mouse zygote, the resulting two cells already have a bias towards differentiation (Piotrowska et al. 2001; Gardner 2001; Zernicka-Goetz 2002). This conclusion contradicts the idea of mammalian embryos beginning life as featureless clusters of cells; moreover, it contributes to understanding the singled-cell zygote as a functional entity.

When the fertilization process is completed, the result is a cell with a specific zygote phenotype, i.e., a new single celled reality that can develop into a complete living organism in a process of self-organization. In the first place, this zygote is more than the mere fusion between female and male pronuclei. Parental genomes are not enough to start the development process because a more complex process of mutual activation of both gametes regulated by cytoplasmic factors in the oocyte is required. In fact, the parental genomes exhibit epigenetic asymmetry at fertilization, differences that are maintained and enhanced in the zygote. The paternal genome undergoes demethylation while the maternal genome undergoes further *de novo* methylation (Surani 2001; Ferguson-Smith and Surani 2001). After the formation of male and female pronuclei, the entry of cytoplasmic factors into the nuclei regulates further epigenetic modifications. This genomic reprogramming allows the formation of a new living organism with a characteristic cellular phenotype, the zygote, which can begin the activation of the development process.

In the second place, the blastocyst, rather than being a symmetrical sphere, is slightly distorted and has recognizable axes (Pearson 2002). The orientation of the first cleavage is related to the position of the polar body and the sperm entry position (Piotrowska 2001; Gray et al. 2004). Moreover, after the first cleavage, blastomeres have distinguishable fates in normal development (Piotrowska 2001). At this two-cell stage, the blastomere that inherited the sperm entry position undertakes its next cleavage division earlier than its sister. This two-cell stage blastomere has a strong tendency to develop into cells that comprise the embryonic parts of the blastocyst (the region containing the inner cell mass) and the later dividing two-cell stage blastomere is destined to develop into the placenta and other supporting tissues (Zernicka-Goetz 2002). Consequently, the first division of the egg influences the fate of each cell and ultimately, the fate of all body tissues. As Zernicka-Goetz points out: "there is a memory of the first cleavage in our life" (quoted in Pearson 2002). The zygote with the asymmetrical and polarized phenotype undergoes a process that involves cell-to-cell communication and co-ordinated cell interaction. Accordingly, the blastomeres are distinct and have different fates, but at the same time their intercellular interaction makes them a two-cell organism.

This aspect limns the difference between a lived cell or a unicellular organism and an embryo from the one-cell stage. All of them will divide into two cells but the former will result in two mutually independent identical cells and the second will produce two distinct cells with different fates, but as part of a development process that will culminate in a complete organism.

APPLICATIONS TO AN ETHICAL DISCUSSION

Scientific advance often goes ahead of the regulation about human embryos research. This imbalance provokes situations where ethical reflection is significantly more difficult and makes normative regulation a real challenge to cover all the possibilities that science presents (Escriba 2002). In this regard, different names have been proposed (e.g., activated cell, clonote, reconstituted egg, or clonocyst) to distinguish the product of SCNT from the embryo resulting from natural or *in vitro* fertilization; consequently different moral or legal status would be attached to each entity (Hansen 2002; Maienschein 2002; McHugh

2004; President's Council on Bioethics 2002). For example, in the UK there was a debate about the Human Fertilization and Embryology Act because it limited its reference to embryos created by fertilization, leaving the status of cloned embryos unregulated (Anonymous 2001). Moreover, a similar ethical debate has been raised in Switzerland in connection with the legal status of embryo-like entities (Burgin 2002).

However, independently of the method used to fertilize an oocyte, the crux lies in whether the zygote phenotype is achieved during the process or not. For this reason, it is important to remember that fertilization is not an instantaneous event but a process that requires specific conditions: a) precise DNA status corresponding to mature gametes (both undergo epigenetic reprogramming) and reciprocal activation of both DNA when gametes are in contact; b) adequate cytoplasmatic environment necessary to activate the molecules used to convey intracellular and intercellular signals; and c) synchronization of both processes (Maleszewski et al. 1999). The conjunction of these three conditions allows for the completion of the fertilization process in the mentioned phenotype zygote.

A variety of situations exist wherein a fertilized egg does not develop correctly into a cell with phenotype zygote, for example when the nuclei is from a somatic cell where epigenetic modifications have not been removed properly (Reik and Dean 2001; Reik, Dean and Walter 2002). This occurred with somatic cell nuclear transfer in an experiment where nuclei from fibroblasts or ovarian cells called cumulus cells were injected in eggs (Cibelli et al. 2001). The results of these tests were not early embryos, but simply a cluster of cells as a consequence of cellular degenerative divisions. The reason for low success in early embryonic development and the presence of aberrations is the failure in the expression of genes due to an incorrect methylation process of DNA. This implies that the necessary synchronization between the transformation of egg-derived and sperm-derived pronuclei and cytoplasmic factors did not occur and, as a result, the cloned cell probably does not have the zygote-phenotype required to begin the development program to form a new living organism.

Notwithstanding, the view that a new diploid cell form from SCNT is different from a zygote, and that it merits a different ethical evaluation is highly problematic mainly for two reasons: one, it could be the case that this new diploid cell from SCNT - or "clonote" - has the same intrinsic potentiality that a zygote created by fertilization; and two, the experiments that try to prove that "clonotes" are radically different from zygotes are not definitive: these experiments are very scarce and have primarily been performed with primates (McHugh 2004). Moreover, the mere use of a different name such as "clonote" seems to be some kind of short cut to avoid confronting the ethical problems posed by this research with human embryos rather than an adequate solution.

Apart from the understanding of the fertilization as a process, the growing evidence of the existence of patterning information in the early mammals embryo has also relevant applications for the ethical discussion. The research in this field has been mainly developed with mouse embryos but there are indications that these recent views may indeed apply to early human development (Denker 2004). Axis determination is present at the zygote stage (and its specification depends on an intrinsic polarity) although it is a fact the plasticity of early embryo and that there are extrinsic factors that can perturb the axis formation of an early embryo. This determination tends to support that the mammalian early embryos are not a uniform ball of cells but a self-organizing organism from the moment of fertilization and can no longer be considered as a mere random collection of cells.

More experiments are also needed as some questions still remain open in order to better understand the mechanisms by which polarity is established (Johnson 2003; Zernicka-Goetz, 2004). However this non-definitive but innovative perspective can considerably enrich future ethical debates about human embryos; particularly this research is challenging the idea of gastrulation as the essential biological process. Scientists are proving that mammals are probably not an exception in the appearance of polarity in the early embryos. Zernicka-Goetz, challenging Lewis Wolpert's statement on the most important time in one's life being not "birth, marriage, or death, but gastrulation", points out that probably patterning of early embryo should be considered "of equal significance to gastrulation or marriage, then if not at least to engagement" (Zernicka-Goetz, 2002). In the same way as that these results compelled to reconsider the scientific evidence about mammals embryo development, we should question the relevance of gastrulation as the reference point in the debate about the moral status of human embryos. These results are proving that human life development between fertilization and gastrulation is something more characteristic of a new entity than a cluster of cells.

As a conclusion, we must recognize that a human embryo is dependent on the successful completion of the fertilization process where a zygote phenotype could develop into a complete living organism in a process of self-organization. Therefore, it is of primary importance to define the terminology to refer to the early embryo in order to avoid it being confusing. Moreover, the conclusions of the debate on the moral and legal status of human embryos should be taken into consideration independently of the method used to create them (the particular debate on each one of these methods is another issue that should be discussed separately). The status conferred to early human embryos would also establish the ethical limits of the research around its fertilization.

In this sense, scientists and bioethicists are central players in the increasingly complex and interdisciplinary negotiations about the rules that should govern this kind of research (Salter 2002). Their responsibility lies, further, in addressing recent research findings as well as their ethical, legal and social implications (Beier 2002; Denker 2004). This involves incorporating new findings on epigenetics and molecular embryology into the ethical debate about embryo research. It is also an important challenge for scientists and scholars in bioethics to make this information accessible and understandable, to nuance public debate on these vital issues.

REFERENCES

Anonymous. 2001. Court backs cloning challenge. *BBC News*, 15 November.<*http://news.bbc.co.uk/1/hi/sci/tech/1657750.stm*> Accessed 23 June 2004.

Beier, H.M. 2002. Der Beginn der menschlichen Entwicklung aus dem Blickwinkel der Embryologie. *Z Arztl Fortbild Qualitatssich* 96(6-7):351-61.

Burgin, M.T., and P. Burkli. 2002. Embryonen und embryo - ahnliche Organismen - Definitionsprobleme im Entwurf zum Embryonenforschungsgesetz. *Ther Umsch* 59(11):613-7.

Bruce DM. 2002. Stem cells, embryos and cloning--unravelling the ethics of a knotty debate. *J Mol Biol* 319(4):917-25.

Check, E. 2002. US biologists wary of move to view embryos as human beings. *Nature* 420(6911):3-4.

Cibelli, J.B., et al. 2001. Somatic cell nuclear transfer in humans: pronuclear and early embryonic development. *E-biomed: the journal of regenerative medicine* 2:25-31.

Denker HW. 2004. Early human development: new data raise important embryological and ethical questions relevant for stem cell research. *Naturwissenschaften* 91:1-21.

Escriba, M.J., et al. 2002. New techniques on embryo manipulation. *J Reprod Immunol* 55:149-61.

Evans, J.P., and H. M. Florman. 2002. The state of the union: the cell biology of fertilization. *Nat Cell Biol* 4 Suppl:S57-63.

Ferguson-Smith, A.C., and M.A. Surani. 2001. Imprinting and the epigenetic asymmetry between parental genomes. *Science* 293(5532):1086-9.

Gardner, R.L. 2001. Specification of embryonic axes begins before cleavage in normal mouse development. *Development* 128(6):839-47.

Gray D, et al. 2004. First cleavage of the mouse embryo responds to change in egg shape at fertilization. *Curr Biol* 14:397-405.

Hansen, J.E. 2002. Embryonic stem cell production through therapeutic cloning has fewer ethical problems than stem cell harvest from surplus IVF embryos. *J Med Ethics* 28:86-8.

Hutchinson M. 2003. Mixed-sex human embryo created. *BBC News*, 3 July. <http://news.bbc.co.uk/2/hi/health/3036458.stm> Accessed 23 June 2004

Johnson, M.H. 2001. Mammalian development: axes in the egg? *Curr Biol.* 11:R281-4.

Johnson, M.H. 2003. So what exactly is the role of the spermatozoon in first cleavage? *Reprod Biomed Online* 6:163-7.

Lauritzen, P., ed. 2001. *Cloning and the future of human embryo research*. New York: Oxford University Press.

McHugh PR. 2004. Zygote and "clonote" – the ethical use of embryonic stem cells. *N Engl J Med* 351:209-11.

Maienschein, J. 2002. Stem cell research: a target article collection: Part II--what's in a name: embryos, clones, and stem cells. *Am J Bioeth* 2:12-9, 30.

Maleszewski, M., et al. 1999. Delayed sperm incorporation into parthenogenetic mouse eggs: sperm nucleus transformation and development of resulting embryos. *Mol Reprod Dev* 54(3):303-10.

Nakao, M. 2001. Epigenetics: interaction of DNA methylation and chromatin. *Gene* 278(1-2):25-31.

Oligny, LL. 2001. Human molecular embryogenesis: an overview. *Pediatr Dev Pathol* 4(4):324-43.

Pearson, H. 2002. Your destiny, from day one. *Nature* 418(6893):14-5.

Pennisi, E. 2001. Behind the scenes of gene expression. *Science* 293(5532):1064-7.

Piotrowska, K., et al. 2001. Blastomeres arising from the first cleavage division have distinguishable fates in normal mouse development. *Development* 128(19):3739-48.

President's Council on Bioethics. 2002. *Human cloning and human dignity: an ethical inquiry*. Washington, DC: President's Council on Bioethics. < http://www.bioethics.gov/reports/cloningreport/index.html> Accessed 23 June 2004.

Reik, W., and W. Dean. 2002. Back to the beginning. *Nature* 420(6912):127.

Reik, W., W. Dean and J. Walter. 2001. Epigenetic reprogramming in mammalian development. *Science* 293(5532):1089-93.

Salter, B., and M. Jones. 2002. Human genetic technologies, European governance and the politics of bioethics. *Nat Rev Genet* 3(10):808-14.

Stolberg, S.G. 2001. Ought we do what we can do? Invoking ethicists to reconcile science with faith. *NY Times (Print)* Aug 12:1WK, 4WK.

Surani, MA. 2001. Reprogramming of genome function through epigenetic inheritance. *Nature* 414(6859):122-8.

Van Speybroeck, L. 2002. From epigenesis to epigenetics: the case of C. H. Waddington. *Ann N Y Acad Sci* 981:61-81.

Vogel, G. 2002. Framework and Stem Cells: The Fight Goes On. *Science* 297(5588):1784.

Wade N. 2002. Stem cells mixing may form a human-mouse hybrid: mice with human cells would be likely. *NY Times* (Print). Nov 27:A21.

Weiss R. 2002. New status for embryo in research. *Washington Post* Oct 30:A01

Zernicka-Goetz, M. 2002. Patterning of the embryo: the first spatial decisions in the life of a mouse. *Development* 129(4):815-29.

Zernicka-Goetz, M. 2004. First cell fate decisions and spatial patterning in the early mouse embryo. *Semin Cell Dev Biol* 15:563-72.

In: Contemporary Ethical Issues
Editor: Albert G. Parkis, pp. 97-112

ISBN 1-59454-536-7
© 2006 Nova Science Publishers, Inc.

Chapter 6

EQUALITY, PRIORITY AND LEVELLING DOWN

Nils Holtug
University of Copenhagen
Department of Philosophy
Njalsgade 80 DK-2300 Copenhagen S Denmark
nhol@hum.ku.dk

ABSTRACT

In this paper, I shall first consider John Broome's account of the difference between egalitarianism and prioritarianism. I shall suggest that while his account plausibly stresses the relational aspect of equality, it does not quite capture its relational nature. Instead, I offer a sketch of an alternative account or, more precisely, I point to some core intuitions about equality I believe that any plausible account of egalitarianism should accommodate. On the basis of this minimal account, I then consider (and criticize) Marc Fleurbaey's claim that there really is no distinction to be made or, more precisely, that prioritarianism is really just a version of egalitarianism

INTRODUCTION

Recently, the distinction between egalitarianism and prioritarianism has received a great deal of attention.[1] Roughly, egalitarianism is the view that the more equally welfare is distributed on individuals in an outcome, the better this outcome is.[2] And roughly, prioritarianism is the view that a benefit to an individual contributes more to the value of an outcome, the worse off the individual is to whom it accrues. Some have argued that a principled distinction can and should be drawn between these two views,[3] others that no difference of any interest can be discerned.[4] Those who believe that there is an important distinction to be made usually claim that unlike prioritarianism, egalitarianism is vulnerable to the levelling down objection. Nevertheless, they differ on whether or not this objection has any real force.[5]

In this paper, I shall first consider John Broome's account of the difference between egalitarianism and prioritarianism.[6] I shall suggest that while his account plausibly stresses the relational aspect of equality, it does not quite capture its relational nature. Instead, I offer a sketch of an alternative account or, more precisely, I point to some core intuitions about equality I believe that any plausible account of egalitarianism should accommodate. On the basis of this minimal account, I then consider (and criticize) Marc Fleurbaey's claim that there really is no distinction to be made or, more precisely, that prioritarianism is really just a version of egalitarianism.[7]

While Broome and Fleurbaey disagree on whether there is an interesting distinction to be made between egalitarianism and prioritarianism, they agree that

---

[1] Arneson (2000), Broome (forthcoming), Brown (2003), Crisp (2003), Fleurbaey (forthcoming), Hausman (forthcoming), Holtug (1998), Holtug (1999), Holtug (2003), Holtug (forthcoming), Jensen (2003), McKerlie (1994), McKerlie (1997), McKerlie (2003), Nagel (1991), Parfit (1991), Persson (2001), Raz (1986), chapter 9, Temkin (1993a), Temkin (1993b), Temkin (2000), Temkin (2003a), Temkin (2003b), Tungodden (2003).
[2] More precisely, this is a characterization of axiological welfare egalitarianism. In the following, I shall only be concerned with axiological principles and with distributions of welfare.
[3] Arneson (2000), Broome (forthcoming), Crisp (2003), Holtug (2003), Holtug (forthcoming), Jensen (2003), McKerlie (1994), Parfit (1991), Persson (2001), Temkin (1993a), Tungodden (2003).
[4] Fleurbaey (forthcoming), Hausman (forthcoming).
[5] Temkin argues that ultimately, the levelling down objection should be rejected (Temkin (1993a)). I criticize his argument in Holtug (1998), Holtug (2003), Holtug (forthcoming), and Temkin responds to my criticism in Temkin (2003c).
[6] Broome (forthcoming).
[7] Fleurbaey (forthcoming).

the levelling down objection does not serve to distinguish them. Against this I finally argue that it is only egalitarianism that is vulnerable to this objection.

THE RELATIONAL NATURE OF EQUALITY

Broome suggests that while both prioritarianism and egalitarianism satisfy the Pigou-Dalton principle of transfer, the former view also generates a strongly separable ordering, whereas the latter view, also, does not.[8] According to the Pigou-Dalton principle, if the sum of welfare remains constant, equality is increased by a transfer of welfare from a better-off person to a worse-off person, as long as their relative positions are not reversed. What this principle captures is the intuition that *equalizing* transfers between two individuals that do not affect the total sum of welfare improve an outcome. Egalitarians and prioritarians share this intuition, but unlike prioritarians, egalitarians are committed to a certain relational value, and this is what Broome attempts to capture by the claim about separability.

A distributive principle generates a strongly separable ordering if and only if it implies that the ordering of the welfare levels of any individual (or subset of individuals) is independent of the welfare levels of others.[9] The claim that egalitarianism does not satisfy this requirement, then, captures the idea that equality concerns the issue of how each individual's welfare relates to that of others.[10] Prioritarians, on the other hand, hold that what matters is an individual's absolute level of welfare, and that the lower this level is, the more priority we should assign to improving it. Prioritarianism, then, does not focus on relations between individuals (although it does of course imply the relational claim that, everything else being equal, it is uniquely best to give a unit of welfare to particular individual if and only if she is worse off than everyone else).

---

[8] Broome (forthcoming), p. 2.

[9] Broome (1991), p. 69.

[10] But interestingly, the lack of strong separability in egalitarianism weakens the case for describing egalitarianism in terms of the Pigou-Dalton principle. This is because we cannot simply assume that if a transfer between two individuals increases equality between them, equality is increased overall. We have to also consider how it affects equality to others. Consider the following two outcomes, where the numbers represent their (three) inhabitant's welfare: A (2,2,4), B (2,3,3). In going from A to B, we transfer one unit from a better off person to a worse off person, while keeping the sum constant. Therefore, the Pigou-Dalton principle implies that B is better. But is it obvious that an egalitarian should make this judgement? While equality increases between the second and the third individual, equality decreases between the first and the second. See also Temkin (1993a), pp. 83-84, and Tungodden (2003), pp. 29-30.

I believe that Broome is correct in holding that prioritarianism generates a strongly separable ordering. I also believe that Broome is correct in focussing on the relational aspect of egalitarianism. Nevertheless, I want to suggest that we can and should be a bit more specific about the nature of this relational aspect (although, as we shall see, this increase in specificity comes at a price).

Egalitarians value equality, and equality is a *relation*.[11] But what does it mean to say that egalitarians *value* this relation? First, they value it *intrinsically*. That is, egalitarians value equality because they take it to be *good in itself* and not (merely) good because it tends to further some other goal, say, fraternity, political stability or the general welfare. Thus, a utilitarian may have reasons to prefer some degree of equality, but surely that does not make her an egalitarian. To be an egalitarian, a person must value equality *for its own sake*.

Second, to qualify as an egalitarian, a person must value equality in the sense that she considers more equal outcomes *in one respect better* than less equal outcomes.[12] However, she need not consider more equal outcomes better all things considered. After all, clearly she may have other concerns, say, for liberty, autonomy or the general welfare.

Combining these two claims, what we get is *outcome egalitarianism*, according to which an outcome is in one respect intrinsically better, the more equal a distribution of individual welfare it includes. Of course, different measures of equality will rank outcomes differently and so there may be genuine disagreements about which outcomes are better than others with respect to this value. Nevertheless, I want to suggest that in order for a measure to be a measure of *equality*, it must imply what we may call *the perfect equality claim*, according to which an outcome in which everyone has the same share of welfare is more equal than an outcome in which individuals have different shares.[13] Like the claim that, according to egalitarians, more equal outcomes are in one respect intrinsically better than less equal outcomes, the perfect equality claim seems to me to be a part of our ordinary conception of what (welfare) equality is.

---

[11] Thus, it is common to stress the relational nature of egalitarianism. For instance, Parfit writes: "Egalitarians are concerned with relativities: with how each person's level compares with the level of other people" (Parfit (1991), p. 23). And Temkin states: "The egalitarian has no intrinsic concern with how much people have, her concern is with how much people have relative to others" (Temkin (1993a), p. 200). Others, however, use the term 'egalitarianism' in a wider sense, where it includes at least one view that does not attach intrinsic value to relations between people, namely prioritarianism (see, e.g., McKerlie (1996), p. 277).

[12] For a similar principle and the suggestion that it is central to egalitarianism, see Tungodden (2003), p. 6.

[13] See also Vallentyne (2000), p. 4.

While outcome egalitarianism and the perfect equality claim each seem innocent enough, together they imply *the egalitarian relational claim*: An outcome in which everyone has the same share of welfare is in one respect intrinsically better than an outcome in which individuals have different shares. The egalitarian relational claim, it seems to me, nicely captures an important relational aspect of egalitarianism. A principle that satisfies it will imply that an increase in an individual's welfare from n to $n+1$ intrinsically improves an outcome in one respect if everyone else is at $n+1$, but makes it intrinsically worse in one respect if everyone else is at n. Thus, the value of such an increase depends on the recipient's welfare level relative to that of others. However, as we shall see, the egalitarian relational claim renders egalitarianism vulnerable to the levelling down objection.

In the following, then, I shall assume that egalitarians are committed to outcome egalitarianism, the perfect equality claim and (hence) the egalitarian relational claim. Of course, these three principles amount only to a very limited characterization of egalitarianism. Thus, they only tell us how to rank equal outcomes against unequal ones and only in one respect. Let me address these two limitations separately.

The proposed egalitarian principles do not tell us anything about how to rank various patterns of inequality. However, they are of course compatible with a number of further restrictions. For simplicity, let us momentarily assume that we hold an all things considered version of outcome egalitarianism (such a version is obtained by deleting 'in one respect' from the original formulation). This version may satisfy, for instance, the Pigou-Dalton principle. But I do not intend to specify the egalitarian ordering of different patterns of inequality here because, as I shall argue in the next section, the characterization I have already provided suffices to distinguish egalitarianism from prioritarianism.

My characterization is also limited in the sense that it does not tell us anything about how to order outcomes *all things considered*. Consider again outcome egalitarianism. Egalitarians will not want to claim that equality is all that matters and so will not simply hold this principle; they will want to combine the concern for equality with certain other distributive concerns. After all, outcome egalitarianism does not even imply that welfare equality at high welfare levels is (in even one respect) better than welfare equality at low welfare levels. Furthermore, if we combine the concern for equality with certain other distributive concerns, there is no guarantee that the resulting principle will imply that an equal distribution is better than an unequal one, all things considered.

Some such principles that combine egalitarian and non-egalitarian concerns will satisfy the pareto principle, according to which two outcomes are equally good if they are equally good for everyone, and one is better than the other if it is better for some and worse for none.[14] Let us call such versions pareto egalitarianism.[15] Because it satisfies the pareto principle, pareto egalitarianism implies that an increase in an individual's welfare (holding the welfare of others constant) improves an outcome even if it changes it from a situation of equality to one of inequality. Furthermore, since pareto egalitarianism is supposed to capture not just the concern for equality but also other relevant moral concerns, I shall take it to be a principle that orders outcomes all things considered. That is, unlike outcome egalitarianism, pareto egalitarianism is a *complete* axiological view (or rather, a class of such views).

Note that while pareto egalitarianism implies that some unequal outcomes are better than some equal outcomes, it nevertheless satisfies the egalitarian relational claim. This is because I have defined pareto egalitarianism as a principle that combines egalitarian and non-egalitarian concerns such as to satisfy the pareto principle, and I have characterized egalitarian concerns in terms of the egalitarian relational claim. Thus, pareto egalitarianism implies that an outcome in which everyone has the same share of welfare is *in one respect* intrinsically better than an outcome in which individuals have different shares.

To put it differently, we might say that egalitarians are committed to a particular *reason* for holding the particular all things considered ordering that they hold. Of course, egalitarians do not share all their reasons (then they would hold the same orderings all things considered), but they have at least *one* reason in common. And this reason is itself an axiological ordering, albeit a partial ordering of outcomes in one respect only.[16]

---

[14]This version differs from standard versions of the pareto principle, since standard versions are formulated in terms of preferences rather than welfare: see, e.g., Broome (1991), p. 152. It corresponds to Broome's principle of personal good: see Broome (1991), p. 165.

[15]While many egalitarians seem to accept the pareto principle, not all do. Brian Barry reconstructs (and endorses) an argument of Rawls', according to which equality is unjust when it is pareto-inferior and, in particular, when there is a pareto-superior outcome in which the worst off are better off; see Barry (1989), chapter 6. Furthermore, Broome defends the pareto principle and combines it with egalitarian concerns in Broome (1991), chapter 8. Also, Tungodden endorses the pareto principle from within an egalitarian framework; Tungodden, (2003), p. 10. Incidentally, both Broome and Tungodden seem to consider the pareto principle a plausible restriction on egalitarianism in its own right, rather than an implication of a plausible weighing of egalitarian and non-egalitarian (e.g. utilitarian) concerns.
On the other side of this egalitarian divide, McKerlie suggests that there is no particular reason why egalitarians should weight equality and welfare such that the pareto principle is always satisfied; McKerlie (1994), p. 287. And Temkin holds a similar view; Temkin (2003b), p. 80.

[16]This ordering is partial because the egalitarian relational claim implies only that equal outcomes are in one respect intrinsically better than unequal outcomes. Thus, it does not tell us how to rank

An implication of accounting for egalitarianism in terms of reasons is that we may find egalitarians and non-egalitarians endorsing identical orderings all things considered. In fact, we may even find egalitarians and prioritarians in agreement here.[17] The difference, of course, will consist in their reasons for endorsing a particular ordering.

But note that this talk of different reasons for accepting a particular ordering all things considered is in no way mysterious. For instance, several reasons have been given for accepting Rawls' difference principle. One of these reasons appeals to Rawls' contract argument, whereas another appeals only to the 'intuitive' case for giving priority to the worst off in cases where they are not themselves responsible for so being.[18] So do two people who both hold the difference principle, but for each their reason, hold the same view? In a sense yes, and in another sense no. In order to bring out the full extent of their commitments, we shall have to refer to their reasons for holding the difference principle. Likewise, in order to describe the difference between egalitarians and other theorists, we shall have to refer to *their* reasons for ordering outcomes in the manner they do.[19]

On this account of egalitarianism, then, it may not always be possible to determine whether an ordering that a particular person holds is egalitarian or not. This will be so if the person does not commit herself to any particular reasons for holding the ordering in question. Furthermore, turning from the person who holds this ordering to the ordering itself, there may be no determinate answer to the question of whether it is egalitarian or not. This may seem to be a rather impractical implication.

However, first, there is nothing that this characterization of egalitarianism prevents us from saying. Even if we cannot always determine whether a particular ordering is egalitarian, we can explain exactly why this is so and what the implications of this ordering are. We can, for instance, say that it is *compatible* with egalitarianism.

Second, as I have stressed, it seems to me that to be worthy of the predicate 'egalitarian', a principle must involve a commitment to outcome egalitarianism,

---

equal outcomes against other equal outcomes and nor does it tell us how to rank unequal outcomes against other unequal outcomes.

[17] For a similar point, see Fleurbaey (forthcoming), p. 9, Jensen (2003), pp. 101-103, and Tungodden (2003), pp. 30-31.

[18] See Barry (1989), pp. 213-214, and Kymlicka (1990), p. 55.

[19] Incidentally, Fleurbaey seems to acknowledge that egalitarianism and prioritarianism rely on different normative reasons even if these reasons do not translate into any difference in the all things considered ordering (Fleurbaey (forthcoming), pp. 2-6). Broome also acknowledges that these two principles rely on different reasons, but claims that this difference does translate into a difference in the ordering (Broome (forthcoming), p.1).

the perfect equality claim and so the egalitarian relational claim. Therefore, in order to be justified in calling a certain distributive view egalitarian, we must understand it as including a particular (egalitarian) reason for accepting it.

Finally, as I have already mentioned, the feature I have called the egalitarian relational claim is what invites the levelling down objection and so drawing the distinction between egalitarianism and prioritarianism in terms of it ensures that this distinction has real theoretical interest.

Before I end this section, I need to briefly specify how egalitarianism, even on my limited characterization, differs from prioritarianism. Egalitarians intrinsically value a particular relation, namely that of equality. This commits them to the egalitarian relational claim. Prioritarians, on the other hand, are not so committed. I suggest that what the prioritarian ascribes intrinsic value to is the compound state of affairs that consists of the state that a benefit of a certain size befalls an individual and the state that the individual is at a particular welfare level, where this value increases when the benefit increases but decreases when the level of welfare increases. Here, there is no commitment to the intrinsic value of equality or to the egalitarian relational claim.

EQUALITY AND PRIORITY

Fleurbaey argues that there is no principled distinction to be made between egalitarianism and prioritarianism. He considers what he calls the minimal egalitarian statement that "unequal distributions have something bad that equal distributions do not have",[20] but claims that this statement has almost no implications for the social ranking. And he considers the claim that prioritarianism implies a strongly separable ordering but suggests that even so, this will only make it a special case of egalitarianism.[21]

Obviously, what Fleurbaey calls the minimal egalitarian statement is very similar to the egalitarian relational claim. In fact, the egalitarian relational claim can be considered an interpretation of Fleurbaey's minimal statement. So what, exactly, is the problem with this statement supposed to be? Fleurbaey stresses that "it is important to distinguish disagreements about the social ranking form disagreements about the *reasons* supporting the social ranking. Only the former have practical implications and are directly relevant for the policy-maker".[22] In

---

[20] Fleurbaey (forthcoming), p. 3.
[21] Fleurbaey (forthcoming), p. 5.
[22] Fleurbaey (forthcoming), p. 2.

other words, what is important for the policy-maker is the social ranking and the minimal egalitarian statement says almost nothing about that.

Let me make two points. First, the egalitarian relational claim *is* defined in terms of a ranking of outcomes, although egalitarians are not committed to this ranking *all things considered* but only *in one respect*. Nevertheless, I am sure that Fleurbaey will insist that what is important for the policy-maker is the all things considered ordering. This brings me to my second point. There is of course a sense in which it is true that what has practical implications for people is the all things considered ordering. What should be implemented is the all things considered ordering and in its implementation, the implications of each separate evaluative element that led to this ordering are not felt by anyone.

But how is this supposed to show that the distinctions between distributive principles should be drawn (only) in terms of such orderings? The policy-maker, or anyone else for that matter, may want to know why she should implement a particular ordering, or whether an ordering that already is implemented is justified, and to answer these questions, she will need to invoke reasons. So distinctions in terms of reasons certainly have real importance.

Fleurbaey further stresses that not only egalitarians, but also prioritarians attach value to equality.[23] More precisely, assuming that the resulting ranking of outcomes is continuous, a prioritarian view can be represented by a function of the form:

$$W = T \times (1-IN)$$

where T is the total sum of welfare, and IN is an inequality index that satisfies the Pigou-Dalton principle. Fleurbaey provides the following proof. We choose W as a function that represents the prioritarian ordering of outcomes and implies that whenever an outcome holds an equal distribution of welfare, W equals T. According to Fleurbaey, it can be shown that such a function exists.[24] We then adopt the following inequality index:

$$IN = 1-(W/T)$$

This is held to be a reasonable inequality index because, first, IN=0 when the distribution is equal. And second, IN satisfies the Pigou-Dalton principle. After all, a prioritarian view satisfies this principle and so any Pigou-Dalton transfer

---

[23] Fleurbaey (forthcoming), p. 6-7.
[24] Fleurbaey (forthcoming), p. 7.

increases W (while T is held constant) and so decreases IN. The point, of course, is that by reversing IN, we get W = T x (1-IN), which is what had to be proved.

In other words, just as pareto egalitarianism can be divided into parts some of which are egalitarian and others are not, so can prioritarianism. However, the egalitarian value to which Fleurbaey thinks the prioritarian is committed is instrumental, not intrinsic.[25] More precisely, Fleurbaey thinks the prioritarian is committed to the instrumental value of equality in a logical sense of 'instrumental value'. Prioritarians hold equality to be instrumentally good in that an equal distribution of welfare is held to be intrinsically better than an unequal distribution of the same sum. This logical sense of 'instrumental value' should be distinguished from its more ordinary causal sense, where an entity has the later sort of value in so far as it causally produces some (intrinsic) value.

In fact, prioritarians must hold that equality has instrumental value not only in the logical sense, but also in the causal sense. Equality has instrumental value in the causal sense because at least certain ways of increasing equality - giving to the worse off rather than the better off - will promote intrinsic prioritarian value.

But even if prioritarians are committed to the instrumental value of equality in both these senses, obviously this does not make them *egalitarians*. Egalitarians hold that equality has *intrinsic* value. As we shall see, this implies (amongst other things) that egalitarians consider levelling down to be in one respect intrinsically good. Prioritarianism has no such implication.

LEVELLING DOWN

While Broome and Fleurbaey differ on whether there is an important distinction to be made between egalitarianism and prioritarianism, they agree that the levelling down objection does not serve to distinguish between them.[26] I now want to argue that, on my construal of these principles, this objection does just that.

According to the egalitarian relational claim, an outcome in which everyone has an equal share of welfare is in one respect intrinsically better than an outcome in which people have unequal shares. Therefore, it is in one respect intrinsically better if inequality is eliminated, even if it does not involve making the worse off better off, but only making the better off worse off. Consider outcomes A and B, where the numbers represent individual welfare levels:

---

[25] Fleurbaey (forthcoming), p. 6.
[26] Broome (forthcoming), p. 2, Fleurbaey (forthcoming), p. 13.

A: (2,1)

B: (1,1)

The egalitarian relational claim implies that B is in one respect intrinsically better than A. According to pareto egalitarianism, such an instance of levelling down does not improve an outcome all things considered. But since, by definition, pareto egalitarianism satisfies the egalitarian relational claim, it implies that such an elimination of inequality, that harms some and benefits none, is *in one respect* intrinsically better. B is intrinsically better than A regarding equality. But how can B be in any respect better? It would benefit no one, not even the worse off. This is the levelling down objection.[27]

What about prioritarianism then? I have already pointed out that there is nothing in the prioritarian's value-commitments that invites this objection. Nevertheless, this claim may be contested. Consider again W, which represents a prioritarian ordering of outcomes. Fleurbaey argues that when a distributional change implies that W is reduced in spite of a decrease in IN, the prioritarian must hold that the change worsens things *in spite* the fact that something good happens on the IN side. Thus, Fleurbaey concludes that "insofar as prioritarians give instrumental value to equality, as shown above, they should also be subject to *a similar kind of criticism* [the levelling down objection]" (my emphasis).[28]

Presumably, the sort of case Fleurbaey has in mind here is a case of levelling down. So let us suppose that the welfare level of the better off is reduced to the level of the worst off, resulting in decreases in W, T and IN. T is reduced because there is a decrease in the sum of welfare and W is reduced because there is a decrease in the sum of *weighted* welfare. And since there is now perfect equality, IN reduces to zero. In other words, levelling down improves an outcome in one respect, namely regarding IN.

However, the *mere fact* that prioritarianism can be split into separate components such as to give rise to an extensionally equivalent ordering does not show that prioritarianism, or anyone who holds it, attaches any relevant kind of value to these separate components. And so the fact that IN decreases in the case described above does not imply that prioritarians are committed to the claim that there is, in any interesting sense, something good about levelling down.

It may be helpful to consider the following analogy. Total utilitarianism can be represented by:

---

[27] Parfit (1991), p. 17.
[28] Fleurbaey (forthcoming), p. 13.

$$U = w_1 + w_2 + \ldots + w_n$$

Furthermore, U is equivalent to the following function of average welfare:

$$V = n \times AVE,$$

where

$$AVE = (w_1 + w_2 + \ldots + w_n)/n$$

Does this mean that total utilitarians are committed to the claim that an outcome that has a higher average welfare than another is in one respect better? Obviously not. Consider the following two outcomes:

C: (-10,*)

D: (-10,-9)

In C, there is only one individual and he has a horrible life (the asterisk represents the nonexistence of a second individual). In fact, he has a negative welfare level. In D, there is also another individual, and she has a life just slightly less horrible, but still clearly on the negative side.[29] Importantly, the move from C to D decreases V in spite of the fact that AVE increases (people are, on average, slightly better off in D). So on Fleurbaey's line of reasoning, total utilitarians must admit that while V decreases, something good is happening on the AVE side.

But obviously, total utilitarians do not consider D to be in any respect better than C. They will hold that D is not intrinsically better in any respect. Furthermore, in such cases, AVE does not have instrumental value in the causal sense. While certain ways of increasing AVE will also increase total welfare, the move from C to D does not do that. What about instrumental value in the logical sense, then? Admittedly, total utilitarians will take AVE to have instrumental value in the sense that, holding n constant, an increase in AVE increases the value of V. But this just means that if we held the number of people in C constant, any increase in AVE would intrinsically improve the outcome. And surely, the mere fact that there are such *other* outcomes that are intrinsically better than C does not imply that, in any interesting sense at least, D is better than C.

---

[29] This case is inspired by a very similar case described by Parfit in Parfit (1984), p. 393.

Likewise, prioritarians do not consider B to be in any respect better than A. As we have seen, they will hold that B is not intrinsically better in any respect. Furthermore, in levelling down cases, decreases in IN do not even have instrumental value in the causal sense. While certain ways of decreasing IN will also increase prioritarian intrinsic value, the move from A to B does not do that. Of course, prioritarians will take decreases in IN to have instrumental value in the (logical) sense that, holding T constant, such a decrease increases the value of W. But this just means that if we held T constant, any decrease in IN would intrinsically improve the outcome. And again, the mere fact that there are such *other* outcomes that are intrinsically better than A does not imply that, in any interesting sense, B is better than A. A prioritarian will happily admit that there are such better outcomes but surely this does not weaken her position. In any case, it is a far cry from claiming that B is intrinsically better than A, which is what the egalitarian claims. And so the 'similar criticism' Fleurbaey believes prioritarianism to be vulnerable to is really not similar at all.

WHAT IS THE DISAGREEMENT REALLY ABOUT?

For all I know, Broome and Fleurbaey may agree that if egalitarianism and prioritarianism are characterized along the lines suggested here, then they have the implications I claim that they have.[30] And I certainly agree that if these principles are characterized along the lines suggested by either Broome or Fleurbaey, then the levelling down objection does not serve to distinguish them. So what is the disagreement really about? Is it, as it turns out, merely a verbal disagreement about how to use certain words?

Although I suspect it includes an element of that, there is more to it. Fleurbaey stresses that it is only the (all things considered) ranking of outcomes that has practical implications and is directly relevant for the policy maker. I, on the other hand, have argued that since the justification of a social policy is also very important, Fleurbaey has not provided a reason for defining distributive principles (only) in terms of overall orderings. Furthermore, I have suggested that the egalitarian relational claim is at the very core of egalitarianism. By saying that it is at the very core I mean that it is an important part of the common sense theory of egalitarianism and an important part of what has attracted many people to this

---

[30] But let me add the following qualification. When criticising Temkin's use of the so-called 'slogan', Broome expresses certain reservations about the clarity of the idea that an outcome can be better with respect to equality but not all things considered better than another (Broome (1991), p. 184). I suspect that his worries carry over to my characterization of egalitarianism.

doctrine. Finally, I have pointed out that my characterization renders the distinction between egalitarianism and prioritarianism theoretically important, since it leaves the former principle but not the latter vulnerable to the levelling down objection. And I take it that Broome and Fleurbaey will disagree with at least some of these claims.

Nevertheless, the disagreement may on some points be less severe than it initially looks. If, for instance, it is agreed that certain (egalitarian) *reasons* for preferring particular rankings are vulnerable to the levelling down objection,[31] it does not matter much whether we *define* egalitarianism in terms of such reasons.[32]

REFERENCES

Arneson, Richard J., (2000), "Luck Egalitarianism and Prioritarianism", *Ethics*, vol. 110.
Barry, Brian, (1989), *Theories of Justice*, London: Harvester-Wheatsheaf.
Broome, John, (1991), *Weighing Goods*, Oxford: Basil Blackwell.
Broome, John, (forthcoming), "Equality versus Priority: A Useful Distinction", in Daniel Wickler and Christopher J.L. Murray (eds.), *"Goodness" and "Fairness": Ethical Issues in Health Resource Allocation* (World Health Organization).
Brown, Cambell, (2003), "Giving Up Levelling Down", *Economics and Philosophy*, vol. 19.
Crisp, Roger, (2003), "Equality, Priority, and Compassion", *Ethics*, vol. 113.
Fleurbaey, Marc, (forthcoming), "Equality versus Priority: How Relevant is the Distinction?", in Daniel Wickler and Christopher J.L. Murray (eds.), *"Goodness" and "Fairness": Ethical Issues in Health Resource Allocation* (World Health Organization).
Hausman, Daniel M., (forthcoming), "Equality versus Priority: A Badly Misleading Distinction", in Daniel Wickler and Christopher J.L. Murray (eds.), *"Goodness" and "Fairness": Ethical Issues in Health Resource Allocation* (World Health Organization).
Holtug, Nils, (1998), "Egalitarianism and the Levelling Down Objection", *Analysis*, vol. 58.
Holtug, Nils, (1999), "Utility, Priority and Possible People", *Utilitas*, vol. 11.

---

[31] By saying that these reasons are vulnerable to this objection, I do not imply that this objection is a good one. That is a separate issue.
[32] I would like to thank Karsten Klint Jensen for some helpful comments on an earlier version of this article.

Holtug, Nils, (2003), "Good for Whom?", *Theoria* vol. lxix.
Holtug, Nils, (forthcoming), "Prioritarianism", in Nils Holtug and Kasper Lippert-Rasmussen (eds.), *Egalitarianism. New Essays on the Nature and Value of Equality*, Oxford: Oxford University Press.
Jensen, Karsten Klint, (2003), "What is the Difference between (Moderate) Egalitarianism and Prioritarianism?", *Economics and Philosophy*, vol. 19 (2003).
Kymlicka, Will, (1990), *Contemporary Political Philosophy*, Oxford: Clarendon Press.
McKerlie, Dennis, (1994), "Equality and Priority", *Utilitas*, vol. 6.
McKerlie, Dennis, (1997), "Priority and Time", *Canadian Journal of Philosophy*, vol. 27.
McKerlie, Dennis , (2003), "Understanding Egalitarianism", *Economics and Philosophy*, vol. 19.
Nagel, Thomas, (1991), *Equality and Priority*, New York: Oxford University Press.
Parfit, Derek, (1984), *Reasons and Persons*, Oxford: Clarendon Press.
Parfit, Derek, (1991), *Equality or Priority?*, The Lindley Lecture, University of Kansas.
Parfit, Derek, (1998), "Equality and Priority", in Andrew Mason (ed.), *Ideals of Equality*, Oxford: Blackwell Publishers.
Persson, Ingmar, (2001), "Equality, Priority and Person-affecting Value", *Ethical Theory and Moral Practice*, vol. 4.
Raz, Joseph, (1996), *The Morality of Freedom*, Oxford: Clarendon Press.
Temkin, Larry S., (1993a), *Inequality*, New York: Oxford University Press.
Temkin, Larry S., (1993b), "Harmful Goods, Harmless Bads", in R.G. Frey and Christopher W. Morris (eds), *Value, Welfare, and Morality*, Cambridge: Cambridge University Press.
Temkin, Larry, (2000), "Equality, Priority, and the Levelling Down Objection", in Matthew Clayton and Andrew Williams (eds.), *The Ideal of Equality*, London: MacMillan Press Ltd.
Temkin, Larry S., (2003a), "Egalitarianism Defended", *Ethics*, vol. 113.
Temkin, Larry S., (2003b) "Equality, Priority or What?", *Economics and Philosophy*, vol. 19.
Temkin, Larry, (2003c), "Personal versus Impersonal Principles: Reconsidering the Slogan", *Theoria*, vol. lxix.
Tungodden, Bertil, (2003), "The Value of Equality", *Economics and Philosophy*, vol. 19.

Vallentyne, Peter, (2000), "Equality, Efficiency, and the Priority of the Worse Off", *Economics and Philosophy*, vol. 16.

In: Contemporary Ethical Issues
Editor: Albert G. Parkis, pp. 113-127

ISBN 1-59454-536-7
© 2006 Nova Science Publishers, Inc.

Chapter 7

A 'SPECIAL' CONTEXT?: IDENTIFYING THE PROFESSIONAL VALUES ASSOCIATED WITH TEACHING IN HIGHER EDUCATION

Bruce Macfarlane
Head of Educational Development and Professor of Education, Thames Valley University, London, UK

Roger Ottewill
Educational Developer, Centre for Learning and Teaching, University of Southampton, Highfield, Southampton, UK SO 17 1BJ. Tel: +44 (0)23 8059 4472 Fax: +44 (0)23 8059 2651 Email: rmo2@soton.ac.uk

ABSTRACT

The paper draws on the philosophy of higher education and existing codes of professional values as a basis for analysing the distinctive ethical challenges of teaching in higher education. For those who teach in higher education there are many values that they share with colleagues in schools and colleges, including respect for learners, collegiality, scholarship and a commitment to reflective practice. Additionally, however, they face a number of ethical challenges that, to some extent, distinguish them from teachers in other settings. These include protecting the academic freedom of students stemming from the goal of promoting student criticality; ensuring respect for learners derived from the concept of adulthood and the principle of andragogy; and accommodating a series of 'dual' roles which define academic identity. While the character of these challenges may vary between countries, arguably they are of international concern in higher education forming a distinctive basis for the identification of universal, professional values.

INTRODUCTION

Attempts to develop ethical standards connected with 'professional' teaching practice in higher education have resulted in recognition that 'values' form an integral part of any definition. In the UK it was the Staff and Educational Development Association (SEDA) that took the first steps in this direction. Formed in 1993, SEDA began accrediting courses in teaching and learning for academic staff in higher education in the same year. Its accreditation framework included a set of 'underpinning principles and values' for informing the practice of higher education teachers (see figure 1).

- An understanding of how students learn
- A concern for students' development
- A commitment to scholarship
- A commitment to work with and learn from colleagues
- The practising of equal opportunities
- Continuing reflection on professional practice

Figure 1. Staff and Educational Development Association Underpinning Principles and Values. Source: http://www.seda.ac.uk/pdf/index.htm

This initiative was complemented by the Dearing Report on higher education (NCIHE, 1997) which recommended that the accreditation of programmes for higher education teachers should be taken forward by a new professional body, the Institute for Learning and Teaching in Higher Education (ILTHE). Established in 1999, the ILTHE produced a statement of professional values (see figure 2) together with an identification of core knowledge areas. The professional values statement was modelled very closely on the SEDA principles and values.

Although individual higher education teachers applying to join the ILTHE were required to provide a reflective statement based on five (later six) core knowledge areas without directly addressing the values statement, higher education institutions seeking accreditation of their programmes had to map the ILTHE professional values against their curriculum learning outcomes. By 2004, it was estimated that some 90 per cent of larger UK institutions had at least one ILTHE accredited programme for new teaching staff (Universities UK, 2004). In January, 2003, the UK government published a white paper on the future of higher education which, *inter alia*, expressed the expectation that all new teaching staff should obtain a teaching qualification incorporating agreed professional teaching standards by 2006 (Department for Education and Skills, 2003). At the time of writing, a consultation process, overseen by the recently created Higher Education Academy (incorporating the ILTHE and other agencies concerned with teaching support and development), is in train to establish an agreed national standards framework for implementation by the beginning of the 2005-2006 academic year (Universities UK, 2004).

- A commitment to scholarship in teaching, both generally and within their own discipline
- Respect for individual learners and for their development and empowerment
- A commitment to the development of learning communities, including students, teachers and all those engaged in learning support
- A commitment to encouraging participation in higher education and to equality of educational opportunity
- A commitment to continued reflection and evaluation and consequent improvement of their own practice

Figure 2. Institute for Learning and Teaching in Higher Education Statement of Professional Values. Source: *http://www.ilt.ac.uk/downloads/040430_ AP_ IERform. doc*

- Reflective practice and scholarship
- Collegiality and collaboration
- The centrality of learning and learner autonomy
- Entitlement, equality and inclusiveness

Figure 3. Further Education National Training Organisation Values. Source: *http://www.fento.org/staff_dev/teach_stan.pdf*

Those awarded Qualified Teacher Status must understand and uphold the professional code of the General Teaching Council by demonstrating that they:

1.1. Have high expectations of all pupils, value and respect their diverse cultural, religious and ethnic backgrounds, and are committed to raising their educational achievement;
1.2. Treat pupils with respect, consistency and consideration, showing awareness of their backgrounds, experience and interests, and having concern for their development as learners more broadly;
1.3. Demonstrate and promote the positive values, attitudes and behaviour that they expect from their pupils;
1.4. Have the ability to communicate sensitively and effectively with parents and carers, and to recognise parents' and carers' roles in pupils' learning, and their rights, responsibilities and interests;
1.5. Can contribute to, and share responsibility in, the corporate life of the schools in which they are trained;
1.6. Understand and support the roles of other professionals in pupils' lives;
1.7. Have the motivation and ability to take increasing responsibility for their own professional development;
1.8. Are able to improve their own teaching, by evaluating it, by learning from the effective practice of others and by using research, inspection and other evidence;
1.9. Are aware of the legal framework relating to teachers' employment and conduct.

Figure 4. General Teaching Council Professional Values and Practice. Source: *http://www.tta.gov.uk/php/read.php?sectionid=110 @articleid =459*

The identification of a set of professional values associated with teaching is not confined to the UK higher education sector. The Further Education National Training Organisation (FENTO) has statutory responsibility for developing standards for teaching in the further education sector. It has issued a set of values of remarkable brevity (see figure 3). Similarly, the Teacher Training Agency oversees standards for the award of qualified teacher status in the compulsory school sector on behalf of the General Teaching Council. This also includes a statement of Professional Values and Practice (see figure 4).

The existence of these separate sets of professional value statements for each educational sector in the UK raises the question of whether there are distinct moral and ethical concerns facing teachers working in higher education. Are there values and moral duties distinctive to teaching in a university context or are the ethical challenges facing educators largely interchangeable?

It is noteworthy, however, that the various statements of values issued by the professional bodies and government agencies concerned with professional teaching practice in UK education appear to share a good deal of common ground, notably with regard to respect for learners, the importance of collegiality and inter-professional working, scholarship and reflective practice (see figure 5). While there are differences of emphasis and tone, the principles espoused in all three statements appear broadly similar in character.

	Respect	Collegiality	Scholarship	Reflective practice
Higher Education Academy	'*respect for individual learners...*'	'a commitment to the development of learning communities...'	'a commitment to *scholarship in teaching...*'	'a commitment to continued *reflection and evaluation...*'
Further Education NTO	'*Entitlement, equality and inclusiveness*'	'*Collegiality* and collaboration'	'Reflective practice and *scholarship*'	'*Reflective practice* and scholarship'
General Teaching Council	'*treat pupils with respect,* consistency and consideration...'	'*understand and support the roles of other professionals...*'	'*...learning* from the effective practice of others and *by using research, inspection and other evidence*'	'*...improve their own teaching, by evaluating it...*'

Figure 5. An analysis of common ground in professional values

This begs the question of what, if anything, is 'special' about the ethics of teaching in a higher education setting? The answer to this question is important in determining the nature of professionalism in higher education practice.

In considering what is distinctive about the ethical challenges of teaching in a university setting, it is important to draw on the wider literature about the 'specialness' of higher education itself. In many ways, the ethics of university teaching is a microcosm of this debate. This literature needs to be related to the value-based statements issued by leading professional and academic societies not only in the UK but also elsewhere, such as the American Association of University Professors and the Canadian Society for Teaching and Learning in Higher Education. In this paper, particular attention is given to three areas where a case can be made for claiming that the ethics of teaching in higher education is 'special' vis-à-vis teaching in other contexts. These areas are the academic freedom of students stemming from the goal of promoting student criticality; the importance of respect for learners derived from the concept of adulthood and the principle of andragogy; and a series of 'dual' roles that define academic identity.

STUDENT ACADEMIC FREEDOM

The claim that higher education represents something philosophically different from other stages in the educational process, particularly compulsory schooling, rests heavily on the emancipatory tradition of the universities in the western world. Barnett (1990) argues that, following in the footsteps of Newman, Jaspers and Habermas, higher education is a liberating process that helps students become independent and critical learners.

> An educational process can be termed higher education when the student is carried on to levels of reasoning which make possible critical reflection on his or her experiences, whether consisting of propositional knowledge or of knowledge through action.
> (Barnett, 1990, p 202)

The role of higher education in enabling students to become 'critical reflectors' both with regard to their own discipline and the world around them is widely understood and supported by western faculty as a teaching objective (Nixon, 1996; Kolitch and Dean, 1999). Ashby's (1969) 'hippocratic oath' for the university teaching profession includes the appeal 'to teach in such a way that the pupil learns the discipline of dissent' (p. 64). This sentiment is reflected in the American Association of University Professors' 'Statement on Professional Ethics' that higher education teachers should protect the academic freedom of students. The statement, which dates back to a declaration originally adopted in 1966, incorporates a number of expectations with respect to professional conduct.

In relation to teaching, the statement makes clear that the first duty of professors is to 'encourage the free pursuit of learning in their students' and to protect their intellectual or 'academic freedom' (AAUP, 1987). While statements concerning teaching values in the compulsory school sector tend to emphasise the importance of respect for the nature of knowledge (e.g. Tomlinson and Little, 2000) they do not reflect the importance attached in higher education to encouraging student critique.

Protecting student (not just staff) academic freedom is a pre-requisite for empowering learners to become critical about knowledge claims. This has a number of implications for higher education institutions and university teachers. While it is easy to state a commitment to student academic freedom certain conditions need to be met to ensure that this principle operates in practice. The management of classroom debate and discussion can often serve as a practical expression of protecting student academic freedom. This demands that the learning environment is a 'neutral and open forum for debate' (Barnett, 1990, p. 8) and that the oral contributions of students are treated with mutual respect (Sachs, 2000). At the same time, there need to be 'ground rules' prohibiting extreme expressions of religious, racial and sexual intolerance especially when debating 'touchy subjects' (Poe, 2000) or 'controversial topics' (Lusk and Weinberg, 1994).

This, it might be argued, is a fairly standard set of expectations relating to the promotion of free and fair discussion in any educational context. However, the particular importance attached to developing critical thinking skills and safeguarding student academic freedom in higher education demand that the teacher's ideological and theoretical dispositions need to be kept under 'restraint' (Macfarlane, 2004). On the one hand, a combination of intellectual honesty and emotional 'leakage' makes it inappropriate and impractical for a teacher to hide their convictions from their students. Being passionate about one's subject is often recognised as a feature of 'good' teaching (Ramsden and Entwistle, 1981), but this can also demand that personal convictions are revealed rather than concealed. On the other hand, research has shown that students expect to receive lower grades if they disagree with their teacher in class (Lusk and Weinberg, 1994), which also poses the more insidious possibility that students will self-censor their work (Macfarlane, 2004). Ensuring what Rodabaugh (1996) refers to as 'interactional fairness' is a complex matter that needs to be sensitive to the essential power imbalance between the teacher and the student. In the Canadian Society for Teaching and Learning in Higher Education's 'Ethical Principles in University Teaching', principle three deals directly with 'sensitive topics'. It recommends that teachers identify their own perspective on the issue under discussion and compare this with alternative approaches or interpretations (STLHE, 1996). This conveys to students the complexity of the issue and the 'difficulty of achieving an 'objective' conclusion'. The university academic must maintain a difficult balancing act. They have a duty to defend their own academic freedom of expression and, at the same time, be concerned to protect the student voice (Evans, 1999).

ADULTHOOD

In most national contexts students in higher education are generally regarded as *de facto* or *de jure* 'adults'. The status of the learner as an adult has practical and moral implications for their involvement in the educational process. As adults, learners are, at least in theory, volunteers, not conscripts legally obliged to attend a period of compulsory education. This means that, unlike school instructors, higher education teachers are not normally seen to be acting *in loco parentis*. Consequently they do not assume the same responsibilities as parents. One effect of this difference is that it places a greater onus of responsibility on students for their own actions while, at the same time, building in higher expectations with respect to confidentiality in, for example, maintaining more restricted access to records of academic progress. In practical terms it means that university teachers do not normally discuss the academic progress of students or matters of a personal kind with parents or anyone else without their permission.

There are, though, threats to the confidentiality of the staff-student relationship. These stem largely from the re-conceptualisation of higher education as a service industry and the status afforded to market-based stakeholders. Parents are seen as one of the 'stakeholders' of modern higher education. This means that some parents may perceive they have a right to be more involved in the university careers of their offspring and to information about their progress. Such a view is particularly likely in the new era of tuition fees in the UK and the probability that, in many instances, parents will be called upon to pay or, at least, contribute to them. In other national systems, tuition fees and parent power are more firmly established.

Amongst academics, parents and students opinions differ as to the desirability of greater parental involvement in the university careers of their sons and daughters. Some see it as disrupting the core relationship between higher education teachers and their students and undermining the transition towards greater personal independence, one of the defining characteristics of adulthood.

Others, however, feel that parental interest and engagement can contribute significantly to helping young people adjust to the demands of higher education.

Whatever stance one adopts on this issue, for higher education teachers there can be a very real tension between, on the one hand, safeguarding the privacy of students and, on the other, responding to reasonable requests for information from parents seeking reassurance as to the progress of their children and that they are obtaining value for money. Many students, however, are likely to take the view that their legal status as adults overrides such considerations.

Along with parents, business organizations are also recognised as important 'stakeholders' in modern higher education. This can raise ethical challenges for the teacher where universities enter into arrangements with business organisations to provide an academic programme (such as an MBA) on a single company basis. Where such agreements exist the essentially voluntary nature of being a higher education student is put at risk by what has been referred to as a 'conscript culture' (Macfarlane, 2000). Refusal to participate can raise fears of being passed over for promotion or even being selected for redundancy. The power of the client organisation as sponsor can, moreover, adversely affect the confidentiality of the staff-student relationship where grades and progress issues are disclosed. The possibility of student academic freedom in classroom discussion, raised in the previous section, can also be endangered where criticism of the organisation occurs or is perceived to have taken place in class. Here there is a fear that class peers will act as whistle blowers, reporting critical remarks as acts of corporate disloyalty, thereby curtailing the open and neutral forum envisaged by Barnett (1990).

Lastly, adulthood implies a qualitative difference in the relationship between teachers and students. Here, Knowles' (1984) theory of 'andragogy' is instructive. As adults, learners are self-directed and expect to take more responsibility for their own decisions. This, as Knowles has argued, has implications for the approach and style of those teaching in higher education through, for example, designing learning based on the experiences of students and making it problem rather than purely content-orientated. This is not just a pragmatic response. It is also an ethical response to teaching adults that necessitates respect for their previous life experiences through building learning which is relevant to their career and personal life. While respect for learners is a universal maxim, it has special resonance for those educating adults.

DUAL ROLES AND RELATIONSHIPS

The nature of higher education is such that academics are often involved in performing dual roles that may not sit easily with each other, thereby giving rise to ethical dilemmas. Three of the most potent dualities are those of teacher and assessor; teacher and intimate; and teacher and researcher.

Teacher and Assessor

The inherent power imbalance between student and teacher in higher education has a number of expressions. One of these is the dual role of teacher and assessor. This, according to Kennedy (1997), is the most significant challenge facing the university teacher. For pupils in compulsory school systems much of the critical summative assessment during the final stages of their school careers involves external examinations where someone other than their teacher acts as assessor. Thus, there is a relatively clear divide between those charged with the provision of learner support (i.e. teachers) and those who assess.

By contrast, in higher education the position is far less clear-cut. Universities, by definition, award their own degrees. This means, in effect, that higher education teachers are often responsible for setting their own summative assessments, both coursework and examinations. Thus, they commonly act as not only teachers but also final arbiters of the performance of their own students.

In performing their assessment role, academics face an in-built tension between the desire to encourage and motivate students to learn and the responsibility to sit in judgement on their performance (Shils, 1982; Kennedy, 1997). This tension is especially acute given the impact that assessment decisions can have on the future career prospects of individual students, whether in academe or more generally. At the heart of any discussion of assessment from an ethical perspective is the virtue of fairness (Macfarlane, 2004). Ultimately, any suggestion that higher education teachers have acted either arbitrarily or inflexibly with regard to assessment will undermine their credibility. Thus, in seeking to address the needs of the individual they have to ensure that they do not act unfairly towards others in the group. While fairness in assessment is an ethical imperative in any stage of the educational process, the comparative autonomy of university teachers makes it distinctive in the higher education context.

Teacher and Intimate

Popular representations of academic life, such as David Mamet's *Oleanna*, often focus on the development of close personal or even sexual relationships between students and teachers. While such relationships can be legally proscribed in the compulsory sector, in higher education, where students are defined as adults and of an age to have intimate sexual relations the situation is more complex. On the face of it students, and in particular mature learners, who enter into close personal and sexual relationships with their teacher do so as fellow, consenting adults. However, given that the relationship between the teacher and the student is fundamentally based on an inequality of power, this poses dangers for the actual and perceived fairness of the educational process, including assessment. Principle 5 of the Canadian Society for Teaching and Learning in Higher Education's 'Ethical Principles in University Teaching' concerns 'dual relationships with students'. While this does not seek to prohibit all close sexual and personal relationships with students, it raises their problematic nature in relation to assessment and supervision duties and suggests that senior colleagues should be notified when such circumstances arise (STLHE, 1996). While such relationships do not automatically imply that assessment practices will be unfair, the perception of favouritism means that they are, at the very least, tainted (Macfarlane, 2004).

Teacher and Researcher

Few university teachers are, by definition, only teachers. Many conduct research and publish as part of their professional role as academics and see it as part of their commitment to the value of scholarship. This dual teacher-researcher role can create a difficult dilemma though for university faculty torn between a desire to carry out research in their field and the time-consuming nature of teaching preparation and supporting student learning via assessment, feedback and tutoring.

Although much has been written about the potential synergy between teaching and research and the need for teaching in higher education to be research-informed, there are tensions especially when the teaching responsibilities of academics do not closely match their research interests. The conditions of mass (or universal) higher education make this increasingly likely, creating a potential cognitive dissonance between the lecturer's teaching duties and personal research goals. Moreover, career reward and recognition structures have historically tended to favour research over teaching excellence. In the UK and Australian context, university teachers have come under increasing pressure in recent years to research and publish as a result of the Research Assessment Exercise (RAE) and Australian Research Quantum. Both are peer-reviewed audits of research excellence resulting in the grading of departments and the differential allocation of research funding from the government on the basis of this grading. This has created greater competitive pressures on institutions to improve their ratings and this, in turn, has had a significant impact on the expectations placed on staff. In such an atmosphere if academics do not fulfil their research potential, for whatever reason, it can be said to cast doubt on their commitment to collegiality.

Here the dilemma is often made more acute by presenting research versus teaching as a 'zero sum game'. In other words, the more time and energy academics devote to research the less they have to devote to their teaching. Inevitably, this can put strains on the quality of the service that they provide for their students. A more productive approach, drawing on the Aristotelian tradition, might be to consider the virtue and associated vices that permeate this dilemma. A critical virtue for both effective research and creative teaching is curiosity. However, an excess of curiosity leads to the vice of obsession, while a dearth can be characterised as a lack of inquisitiveness. To maintain an appropriate balance between research and teaching, lecturers need to hold fast to the virtue of curiosity while resisting the temptation to become obsessive about either. Obsession, particularly where it is directed at one aspect of a researcher/teacher role, is every bit as damaging as being disinterested in either or both.

THE VALUES OF UNIVERSITY TEACHING

In developing professional standards for teaching in higher education, due recognition needs to be given to the distinctive nature of the setting. Arguably, the statements of professional values produced by the American Association of University Professors and the Society for Teaching and Learning in Higher Education in Canada are more finely tuned to the particular circumstances of those teaching in a university environment than those of the UK's Higher Education Academy or Staff and Educational Development Association.

Taking a lead from the American and Canadian statements, the analysis developed in this paper has identified six values, which are of special importance to the professional practice and behaviour of teachers working in higher education (see figure 6).

- Active protection of student academic freedom
- Confidentiality in the student-teacher relationship
- Respect for the prior learning and experiences of students as adults
- Fairness in the exercise of the power of assessment
- Transparency with regard to dual relationships with students
- Managing the tensions of the dual teacher-researcher role

Figure 6. Some values distinctive to university teaching

It might be contended that two of the values presented in figure 6 – fairness in assessment and transparency in dual relationships – are equally applicable to the role of the teacher in other phases of the educational process. However, the first of these stems from the unique dual role of higher education faculty as teachers and assessors while the second demands transparency principally to ensure fairness (and the perception of such) with regard again to assessment. The values presented in figure 6 do not necessarily cover all aspects of teaching in higher education, but they do relate to some of the key distinguishing features of university life. Moreover, they are essentially intended as latter day Aristotelian 'virtues' rather than a detailed and prescriptive code of conduct. They provide an analytical basis for further work in designing a values statement suitable for professional teaching standards both nationally and internationally.

CONCLUSION

As Tomlinson and Little (2000) recognise, university teachers face additional dilemmas to those working in other parts of the education system. It is with the nature of these dilemmas and the extent to which they can be addressed with a framework of values common to all members of the teaching profession that this paper has been primarily concerned. The stance adopted here is that such dilemmas present those teaching in higher education with ethical challenges that can only be partially met by recourse to a common set of values. While the character of these challenges may vary across national boundaries, in most respects they are likely to be of universal concern.

With the globalisation of higher education the need to ensure a degree of harmony between the values statements of individual countries becomes increasingly necessary. At a minimum, where there are differences, academics and students moving between countries need to be made aware of them. However, in line with the foregoing analysis the similarities are likely to be greater than any differences. In the UK context, it can only be hoped that in developing a framework of professional standards for teaching in higher education adequate attention is paid to the special nature of the ethical challenges that face faculty members.

REFERENCES

AAUP: American Association of University Professors' Statement on Professional Ethics. (1987). Retrieved on 6 August, 2004, from *http://www.aaup.org/ statements/Redbook/Rbethics.htm*

Ashby, E. (1969). A hippocratic oath for the academic profession, *Minerva*,8 (1), Reports and Documents, 64-66.

Barnett, R. (1990). *The Idea of a Higher Education*. Buckingham: The Society for Research into Higher Education/Open University Press.

Department for Education and Skills (2003). *The future of higher education: Creating opportunity, releasing potential, achieving excellence*. London: The Stationary Office, London.

Evans, G. (1999). *Calling academia to account: Rights and responsibilities*. Buckingham: SRHE/Open University Press.

FENTO: Further Education National Training Organisation Values (1999). Retrieved on August 2, 2004, from *http://www.fento.org/staff_dev/teach_stan.pdf*

General Teaching Council Professional Values and Practice (n.d.) Retrieved on August 2, 2004, from *http://www.tta.gov.uk/php/read.php?sectionid=110 @articleid =459*

ILTHE: Institute for Learning and Teaching in Higher Education's Statement of Professional Values (1999) Retrieved on August 2, 2004, from *http://www.ilt.ac.uk/ downloads/040430_ AP_ IERform. doc*

Kennedy, B. D. (1997). *Academic duty*. Harvard: Harvard University Press.

Knowles, M. (1984). *Andragogy in action: applying the modern principles of adult learning*. San Francisco: Jossey Bass.

Kolitch, E. & Dean, A V. (1999). Student ratings of instruction in the USA: hidden assumptions and missing conceptions about 'good' teaching. *Studies in Higher Education*, 24 (1), 27-42.

Lusk, A. B. & Weinberg, A S. (1994). Discussing controversial topics in the classroom: Creating a context for learning. *Teaching Sociology*, 22, 301-308.

Macfarlane, B. (2000). Inside the corporate classroom. *Teaching in Higher Education,* 5 (1), 51-60.
Macfarlane, B. (2004). *Teaching with integrity.* London: RoutledgeFalmer.
Nixon, J. (1996). Professional identity and the restructuring of higher education. *Studies in Higher Education,* 21 (1), 5-16.
NCIHE: National Committee of Inquiry into Higher Education (1997). *Higher Education in a Learning Society: Report of the National Committee.* London: HMSO.
Poe, R E. (2000). Hitting a nerve: when touchy subjects come up in class. *Observer,* 13 (9), 18-19, 31
Ramsden, P. & Entwistle, N. J. (1981). Effects of academic departments on students' approaches to studying. *British Journal of Educational Psychology,* 51, 368-383.
Rodabaugh, R (1996) Institutional commitment to fairness in college teaching. In L. Fisch (Ed.), *Ethical Dimensions of College and University Teaching: Understanding and honouring the special relationship between teachers and students* (pp 37-45). San Francisco: Jossey-Bass.
Sachs, J. (2000). The activist professional, *Journal of Educational Change,* 1 (1), 77-95.
Shils, E. (1982). The academic ethic. *Minerva,* 20 (1-2), 107-208.
STLHE: Society for Teaching and Learning in Higher Education's Ethical Principles in University Teaching. (1996). Retrieved on Aug 7, 2004, from *http://www.umanitoba.ca/academic_support/uts/stlhe/ethical.html*
SEDA: Staff and Educational Development Association's Statement of Underpinning Principles and Values. (1993). Retrieved on August 2, 2004, from *http://www.seda.ac. uk/pdf/index.htm*
Tomlinson, J. & Little, V. (2000). A code of the ethical principles underlying teaching as a professional activity. In R.Gardner, J. Cairns & D. Lawton (Eds.) *Education for Values* (pp.147-157). London: Kogan Page.
Universities UK (2004). *Towards a Framework of Professional Standards.* London: Universities UK.

Chapter 8

UTILITARIANISM, REPUGNANT PLEASURES AND MORAL EXPLANATION

Hugh Upton
Philosophy & Health Care
University of Wales Swansea
UK

ABSTRACT

In proposing that actions are made morally right by their production of something held to be good, utilitarians have traditionally had in mind what we can call a natural commodity or state. To call something 'natural' is not of course a philosophically straightforward assertion but the relevant implication of this for our purposes is that it can be identified and characterised without recourse to some prior moral evaluation. Our starting point will be this traditional conception of the theory, and the problem to be considered will be outlined initially in terms of just one of the examples, the production of pleasure. Whilst this choice among natural goods may strike some as tendentious, it is made only for ease of exposition. It will be argued later that the selection is actually immaterial with respect to the issue in question.

I

In proposing that actions are made morally right by their production of something held to be good, utilitarians have traditionally had in mind what we can call a natural commodity or state. To call something 'natural' is not of course a philosophically straightforward assertion (see Michael Smith, 2000, pp.92-96) but the relevant implication of this for our purposes is that it can be identified and characterised without recourse to some prior moral evaluation. Familiar examples have been the experiences of pleasure or of happiness, or the fact of a preference having been satisfied. Our starting point will be this traditional conception of the theory, and the problem to be considered will be outlined initially in terms of just one of the examples, the production of pleasure. Whilst this choice among natural goods may strike some as tendentious, it is made only for ease of exposition. It will be argued later that the selection is actually immaterial with respect to the issue in question. So, in terms of hedonistic utilitarianism, the problem can be stated as follows: the generation of pleasure is what is held to be right-making yet we know and deplore the fact that some people take a sadistic pleasure in the suffering of others. Surely, then, we cannot regard the fact that an act would be productive of such objectionable pleasures as contributing to its moral rightness.

In setting out this criticism something must be said about the methodology to be employed. Most generally, it will be assumed that in constructing and assessing moral theories we are to some degree accountable to our current moral beliefs. This is not to deny, of course, that some of the latter may be abandoned or modified in the process, but merely that any theorising of this kind must regard it as possible, even likely, that some will be accommodated as they stand. More specifically, it will be assumed that there is a widely shared abhorrence of many typical instances of sadistic actions, together with a widely (but, as will be seen, not universally) shared sense of an overwhelming incongruity in supposing that the perpetrators' pleasures could count towards the moral rightness of these actions. It is important to stress, though, that neither of these more specific assumptions necessarily involves the claim (which may or may not be defensible) that *all* pleasure taken in the suffering of others is objectionable. To take one example, some may wish to deny this on the grounds that it is acceptable within the context of consenting partners. Whatever our feelings about this, there is no need to reject the possibility here, since it is unlikely to mitigate our concerns about pleasure deriving from the infliction of suffering in the absence of consent. To take another example, it may be thought acceptable to feel pleasure at the suffering imposed as part of the just punishment of those guilty of crime. It may even be argued that the contribution made by the criminal's suffering to the satisfaction of a legitimate grievance felt by the injured party contributes to the moral justification of the act of punishment. Again, there will be no difficulty in our assuming that this is arguable, that such pleasures may at the very least be morally acceptable. With both examples, however, we should note that their acceptability depends upon special moral considerations that do not in any obvious way derive from utilitarianism: the distinctive significance of consent, and the particular circumstances that might render suffering a morally proper object of pleasure through the guilt of the criminal who suffers. Thus, we need not deny that some non-utilitarian theory might show that, in specific circumstances, a certain kind of pleasure in another's suffering could be legitimate. The claim, rather, is that hedonistic utilitarianism is indiscriminate in such matters and is unable to exclude from its conception of what is right-making those pleasures that most will regard as morally abhorrent.

So, whatever our moral assessment of pleasures deriving from the consensual infliction of suffering, or the satisfaction of a legitimate grievance, our concern here will be with the existence of what is (let us assume) plainly objectionable pleasure taken in the suffering of unwilling victims. There is of course a certain amount of variety here, ranging from the form that has some limited social admissibility under the name of *schadenfreude* to the sort of sadism that is perhaps the most repellent of all human attributes. In fact, it may be that even this range should be seen as only part of a wider problem of morally dubious pleasures. Consider someone's pleasure in their skill at cheating others, or in their own virtue, or indeed any pleasure relating to what James Griffin (1986, pp.320-21) engagingly called 'the whole gamut of shabby desires'. However, it is pleasure taken in the suffering of others that poses the problem most forcibly, for it seems that the utilitarian must take even the most repugnant pleasure of the sadistic torturer as a good that morally supports (even where it is insufficient to require) the performance of an act that will produce it. That is, the pleasure must be understood to be what is right-making, what would make the act right were there only to be sufficient of it created. Denying this would leave utilitarianism with what seems a rather implausible position, that (probable) instances of pleasure lack any explanatory moral significance unless they form part of the (probable) maximising outcome, whereupon they fully explain the moral rightness of the act.

The problem can be exhibited equally well in terms of what reasons would have to make moral sense to us. To make use of a concept that was given a valuable elucidation by Shelley Kagan (1989, p.17), we can say that we would have to accept the likelihood of the sadistic pleasure occurring as being a moral reason *pro tanto* for any act that would produce it; a reason, in other words, that invariably has moral weight and which persists as a reason even when overridden by other considerations. Again, the denial of this seems implausible: that probable pleasure is not a moral reason in support of an action unless it forms part of the likely maximal outcome, when it immediately constitutes the sole moral reason for action.

The claim, then, is that the problem of repugnant pleasures is fundamentally a problem not (as it is usually seen) for the account of right action but for that of moral explanation. It is a problem that, while remaining essentially the same, can be revealed in two different forms. For the first of these we can take the case of the torturer, a case where we will assume that, despite the associated pleasure, the act is nevertheless sub-optimal from the utilitarian point of view. Even so, not only will the torturer's pleasure still have to be understood by the utilitarian as right-making but, as Geoffrey Scarre (1996, p.155) has expressed it, his pleasure must be seen as morally offsetting the suffering of the victim. We must suppose, in other words, that at least some moral good came out of the episode. For the other form of the problem, though, we do not have to restrict ourselves to actions that utilitarianism will show to be wrong. Consider a case that falls short of torture but constitutes what might be regarded as legitimately intimidating and aggressive interrogation of known terrorists, where the aim is to prevent the loss of innocent lives. Let us suppose both that utilitarian calculations show this form of interrogation to be most likely to maximise the total of pleasure, and that on other grounds non-utilitarians also regard it as morally justified in the circumstances. Let us suppose also, however, that while the actions of the interrogators do not exceed what is held to be justified on either basis, they derive a sadistic pleasure from the suffering and extreme anxiety that they are able to inflict on their prisoners. Here, it would seem, utilitarians not only have to take a morally dubious pleasure to be part of what is generally morally right-making, but must see it as included in what makes these specific actions morally right.

Once we see the problem in this way it seems likely that we will have to reject one familiar kind of response as failing to address the main issue. We find it, for example, in the work of R.M. Hare. Hare has argued (1981, pp.140-42) that it is unlikely that acts of sadism will maximise the total of pleasure and thereby be made right, and that it would plainly be better in utilitarian terms to seek pleasures that do not involve inflicting suffering upon others. The Romans, as he says, should have given up the cruelties of their arena games in favour of chariot races and football matches. However, attractive though this suggestion may be, it does not touch on the underlying moral problem. That a sadistic act is likely to be sub-optimal in respect of the production of pleasure does not alter the fact that the associated pleasure is morally right-making, and indeed that it may quite easily be a component of a situation that is actually optimal, as in the type of case mentioned above, where the act produces a sadistic pleasure that is merely part of the maximal total, most of which may be pleasure of the most innocent nature. On the same basis it can be argued that while Scarre (1996) has produced an excellent analysis of the problem, one that draws attention to the fact that it does not depend for its force on any sadistic act ever turning out to be maximising, his own proposed solution is not adequate. In broad terms, he argues that utilitarianism can be defended by means of a 'sufficiently subtle theory of human well-being to show why sadistic pursuits are not just other-harming but self-harming too' (Scarre, 1996, p.155). No doubt there are contexts in which such a theory would be highly desirable, but it would seem to be of strictly limited relevance here. At best, I suggest (see Upton, 2000) it would contribute further utilitarian reasons for supposing such pursuits to be sub-optimal, rather than ways of countering his own disquiet over the idea that their characteristic pleasures must be regarded by the utilitarian as morally offsetting any harm that is done.

There thus seems to be a good basis for arguing, quite generally, that we cannot defend hedonistic utilitarianism by trying to show (or even by successfully showing) that no sadistic act will ever be maximising. The counter-intuitive idea that the associated pleasures would still have to be seen as morally right-making would remain. Before moving on, though, we should perhaps consider what help there might be in making reference to the various structures that constitute 'two-level' or 'indirect' utilitarianism. Here we can restrict ourselves to the two broad kinds of proposal that are made under these headings.

First, as Brad Hooker (2000, pp.142-143) notes, it is entirely usual for act-consequentialists to advocate the use of the customary moral rules in reaching our everyday decisions, rather than on each occasion attempting directly to assess the expected utility of each of the actions open to us. As he says, lack of information, shortage of time, the avoidance of personal bias and the preservation of trust in society are all reasons for favouring this as a decision procedure. Suppose, then, in order to judge the theoretical effect of this procedure on our problem, we simply assume that these reasons hold with respect to the characteristic actions of the sadist. Let us take it that the sadist will be expected to follow the rule of (for example) not deliberately causing suffering, rather than trying to calculate whether, on this occasion, some sadistic action might actually maximise human happiness. Has the introduction of this decision procedure solved the problem? I would suggest not, since, once again, we seem only to have addressed the problem of morally repugnant actions, not the dubious status of the pleasures that have to be understood as providing, *pro tanto*, a moral reason for those actions. The introduction of this sort of decision rule into act-consequentialism (and thus into hedonistic act-utilitarianism) may make certain kinds of unpleasant act less likely, but it can have no bearing on the moral plausibility of the value to be promoted. After all, the rules have no status apart from the primary utilitarian principle. They are justified only in so far as following them would increase our success in choosing actions that will maximise (to return to hedonism) the total of pleasure, regardless of its source or nature. Thus it will remain true that ultimately any moral explanation offered by this form of utilitarianism will be based on the presupposition that all instances of pleasure are morally right-making.

The second proposal that we need to consider is the version of utilitarianism that goes beyond this kind of moral pragmatism with regard to rules and takes their construction to be one of the defining elements of the theory. For rule-utilitarianism the rules are in themselves the proper object of moral assessment, rather than merely guides to discovering the right action in circumstances unfavourable to careful calculation. For this reason it may seem to provide a more powerful response to our problem, since the hypothetical sadist is now confronted with a requirement to follow the rule against deliberately causing suffering and is positively denied the option of discarding the rule on the basis of his own calculations, however carefully they might be done. Yet in fact, surely, what we find in this response is a flaw parallel to the one relating to individual acts. Just as act-utilitarians have to take the likely pleasure that will result from an act as invariably counting morally in favour of its performance, so rule-utilitarians have to accept the likely pleasure resulting from following a rule as counting morally in favour of its acceptance. Thus, although the rule 'torture others whenever it amuses you' is most unlikely to be adopted as maximising, the rule-utilitarian is nevertheless committed to seeing some moral reason in favour of its acceptance. Or, to look at the issue another way, any rule designed to protect people from becoming unwilling victims of our sadist must be seen by the utilitarian as having its moral worth *on balance*, to be advocated *despite* the loss of any sadistic pleasures that will be precluded through its observance, since the possibility of such pleasures must be understood as counting morally in favour of abandoning the protective rule; although, *ex hypothesi*, it will not count overwhelmingly. So, far from being removed by the introduction of rules, the counter-intuitive element of utilitarianism seems simply to persist in a slightly different form.

II

We have been looking so far at responses that admit there to be something morally disturbing within utilitarianism, but which claim that it can be defused and accommodated harmlessly within the theory. Having suggested that these fail, we can now turn to attempts to refute the objection in a more direct way, where this involves denying that there is anything that should disturb us in the first place. Certainly there have been a number of writers ready to uphold utilitarianism as a moral theory by taking this line, with Bentham being perhaps both the most celebrated and the most trenchant of them. To be brief with the basis of his theory, we can recall that in *An Introduction to the Principles of Morals and Legislation* Bentham (1967) first defines the principle of utility as 'that principle which approves or disapproves of every action whatsoever, according to the tendency which it appears to have to augment or diminish the happiness of the party whose interest is in question' (p.126). The principle is then asserted to be the only source of meaning for the words 'ought', 'right' and 'wrong', and the claim is further made that a proof of its rectitude is neither possible nor required (pp.127-128). Bentham thereby takes himself to have established that any principle at variance from that of utility must be wrong. Of particular interest among these mistaken principles is the one he calls 'asceticism', a principle which is the 'inverse' of the principle of utility, and thus approves of actions in so far as they diminish happiness and disapproves of them in so far as they increase it (pp.132-133). It is with this principle of asceticism in mind that he then makes the claim that is of particular relevance to our problem, and writes:

> It is only upon that principle, and not from the principle of utility, that the most abominable pleasure which the vilest of malefactors ever reaped from his crime would be reprobated, if it stood alone. The case is, that it never does stand alone; but is necessarily followed by such a quantity of pain (or, what comes to the same thing, such a chance for a certain quantity of pain) that the pleasure in comparison of it, is as nothing: and this is the true and sole, but perfectly sufficient, reason for making it a ground for punishment (p.133).

Support for this position comes from T.L.S. Sprigge, at least with respect to what he calls 'untroubled wickedness' (1988, p.235), which is wickedness that the perpetrator experiences as a sheer pleasure, devoid of feelings of disgrace or shame. Making a contrast with the troubled variety, he writes:

> More cheerful [untroubled] wickedness, however awful in its results for others and however repellent for those troubled at the thought of these results, is good in itself, just as Bentham thought. The pleasures of the sadistic bully go to the credit side of the balance of good and evil in the world, however much they inevitably lead to a much greater weight on the debit side (p.238).

It is also worth noting here the suggestion of J.J.C. Smart (1973) who, in effect, puts forward a thought experiment in support of Bentham's proposal that we should reflect on the pleasure when it is hypothetically separated from the suffering caused by the crime. Consider a universe in which there is just a single sentient being, a sadist who gains enjoyment from his (actually false) belief that others exist and are suffering torments. Compare this with a universe that differs in only a single respect: that the sole sentient being feels sorrow over the same false belief. Given that there is happiness in the first that is absent in the second, and no possibility of harm to others in either, Smart suggests that 'the universe containing the deluded sadist is the preferable one' (p.25).

What we have, then, in various forms, is the claim that the critics of utilitarianism misunderstand the situation through being distracted by the suffering typically caused by the drive for sadistic pleasures. Were they to focus instead on the value of the pleasure in itself they would be compelled to admit that this, like all pleasures, at least adds to the total of good in the world. Thus, we are to suppose, if this line of argument is successful, the objection to conceiving of sadistic pleasures as morally right-making would be removed.

In responding to this, we should first of all just note that an expression such as 'so-called' needs to be taken as read before Bentham's use of the term 'abominable', since on the view we are addressing there may indeed be vile malefactors but strictly there can be no abominable pleasures. A mental state comprising, for example, intense joy at the suffering of a child must be seen as in itself irreproachably good, criticism being appropriate only if someone were actually to cause the suffering in order to obtain the pleasure or (presumably, by extension) where their taste for such pleasures rendered them so likely to act on it in this way that the probable disutility of the taste outweighed its probable value as a source of good.

To assess the plausibility of this claim regarding the status of the pleasure considered in itself we must keep in mind that utilitarianism is being judged as a moral theory, where this is taken to refer to a systematic attempt to provide, or substantially contribute to, a theoretical account of morally right action. Since utilitarianism must therefore meet at least the standard requirements of this kind of construction, it is worth enlarging a little on the significance of the concern over sadistic pleasures in the light of the work expected of a moral theory. Central to this work, surely, is that such a theory must, in some sense, give us an explanation of why our actions are right or wrong, not merely indicate which category they fall into. A specific moral problem may be useful here as an illustration. There is a familiar consequentialist line of argument that tries to justify state punishment by reference to a principle of deterrence: that punishment is justified as an essential contribution to a threat that will dissuade some proportion of potential criminals from criminal action. Equally familiar is the objection that, under certain circumstances, this would require a conspiracy by the authorities deliberately to convict someone believed innocent, simply to achieve the good of the dissuasive example. By way of response it is of course quite legitimate to point out the unlikelihood of such a proposal actually being supported by an assessment of the probable consequences, given the difficulties of securing the necessary secrecy and the great disvalue attendant on the ruse being discovered. Yet even were we to accept that no such case would ever in fact be warranted by the likely consequences, we might still feel that the principle of deterrence was inadequate as a theory of punishment in virtue of its failure to register what is presumably our main concern with any such conspiracy: the unfairness inherent in the act of punishing those believed to be innocent. We have, in other words, a theory that succeeds (let us assume) in ruling out all of a certain class of unjust acts, but which does so on grounds that, far from enlightening us as to the injustice, seem rather to be perverse in their systematic exclusion of what is morally most relevant.

While the possibility (and desirability) of moral theories remains highly controversial, it seems to be generally accepted that some kind of explanatory role is an appropriate part of a test for their adequacy. Here, for example, is an account of the role from the defence of consequentialism by Hooker (2000, p.4) who claims that moral theories 'should identify a fundamental principle that both (a) explains why our more specific considered moral convictions are correct and (b) justifies them from an impartial point of view'. In their work on the application of theoretical ethics to practice, T.L. Beauchamp and J.F. Childress (2001) also choose to distinguish explanation and justification. Under their conception, explanatory power is the provision of 'enough insight to help us understand the moral life: its purpose, its objective or subjective status, how rights are related to obligations, and the like', while justificatory power is the capacity to 'give us grounds for *justified* belief, not merely a reformulation of beliefs we already possess' (p.340, emphasis in original). Martha Nussbaum (2000) is another who has drawn attention to the role of theories in explaining why certain conduct is judged to be right. They should show, she suggests, the 'point' of the various rules, in terms of some more general imperatives, or end, or plurality of ends, such that 'the reason of the agent is addressed with persuasive considerations that illuminate rules of conduct, giving an intelligent being something to go on in deciding whether she wants to adopt the rule in question' (p.237). Plainly there will be many interpretations of the requirement and many ways of phrasing it, but the general idea is clear enough: a moral theory is expected not merely to advise us on what we ought to do, but also, in the light of certain general aims or imperatives, exhibit this advice as morally more intelligible and more defensible than would otherwise have been the case.

Returning to the question of the place of sadistic pleasures in a theory of hedonistic utilitarianism, we might usefully approach the issue in the light of John Rawls's well-known account of 'reflective equilibrium' in moral theory. This part of moral philosophy, Rawls argues, begins with our moral judgements and then attempts to produce 'a set of principles which, when conjoined to our beliefs and knowledge of the circumstances, would lead us to make these judgements *with their supporting reasons* were we to apply these principles conscientiously and intelligently' (Rawls, 1999, p.41, emphasis added). Assuming such a position of equilibrium to be achievable, we are to keep open the possibility that, as we move towards it, adjustments might be made either to our judgements or to the more general moral conception. Thus, having considered various general conceptions, a person would reach equilibrium when 'he has either revised his judgements to accord with one of them or held fast to his initial convictions (and the corresponding conception)' (p.43).

Suppose, then, that we apply Rawls's conception of equilibrium not so much to the question of what is right but to the question of what makes sense to us as moral reasons or explanations. On one side we will have the general conception of hedonistic utilitarianism, holding that pleasurable states of mind are what we desire intrinsically, that they are thus of intrinsic value, that pleasure is thus the good, and that the morally right act is one that maximises its likely production. On the other side we have a number of specific judgements that are in accord with the general conception but morally (I am assuming) highly counter-intuitive. For example, are we likely to say that the moral rightness of the aggressive interrogation is partly explained to us by the pleasures enjoyed by the interrogators? Does this add to our insight into the (or any) moral life, or present us with morally persuasive considerations? Similarly, we can ask the same kind of question in respect of any dubious pleasures that must be supposed to count towards the rightness of an act that would generate them, despite such an act being sub-optimal. Does a theory that regards the sadistic pleasure of the torturer as right-making, as a moral reason *pro tanto* for the act of torturing, thereby help reveal the point of our ordinary moral objection to such actions, or in any way render our rejection of the act morally more intelligible? With such questions, of course, we reach a familiar problem of methodology. Philosophy at this point inevitably relies on facts about people's ordinary moral thinking and it is thus important not to prejudge individual reactions by any unqualified assertion about how 'we' would respond. Let me then merely report my own findings for consideration: that, far from increasing the moral intelligibility of the situation, such proposals tend to be seen as morally bizarre. Even granted that we must allow that a moral theory will not only systematize our moral judgements but may also change some of them, this kind of change would appear to be deeply at variance with any ordinary sense of the point of moral judgements.

For the required comparison, let us go back to the general conception and see whether the attempt to restore equilibrium would require anything of equal or greater implausibility than these putative moral explanations. Minimally, we would have to deny that the fact that a certain kind of mental state was desired for its own sake was a compelling reason to regard its existence as invariably morally good. We would be required, that is, to accept that whilst some kind of mental state may be unconditionally desirable to us, it remains a further question whether it has moral value. While again being aware of the methodological requirement for caution here, I would suggest that such a constraint on our general theorizing has far greater plausibility than the more specific explanations under consideration, and that the move to equilibrium therefore calls for its acceptance. It would seem, then, that for anyone whose response is of this kind, hedonistic utilitarianism must fail to meet the explanatory requirement of a moral theory, even on the (large) assumption that its overall identification of the right actions for us to perform were found acceptable.

III

So far we have been examining the problem with respect to one well-known variety of utilitarianism. We need now to look at the question of whether consequentialist theories more generally can be defended from it and, if so, by what means. To begin, we can ask whether the initial selection of pleasure as the good has been prejudicial to the discussion. Can the problem exemplified by repugnant pleasures be avoided by modifying the account of the utilitarian (or, more generally, consequentialist) good? One point to note at the outset is that avoidance of the problem will require that there is no *unqualified* reference to pleasure at all, since so long as its unrestricted form remains as even part of the specified good there is the potential for its repugnant instances to make their implausible appearances in our moral explanations. Such a requirement is already quite far-reaching in its implications, since while most consequentialists do not favour a purely hedonistic form of the theory, many will wish to include pleasure without qualification as one component of the good, whether (see Hooker, 2000, pp.37-43) in terms of an 'objective list' account or in terms of a direct appeal to the value of certain subjective mental states. Yet simply adding further goods will not remove the problem of this one, and nor will the substitution of a merely more complex conception of pleasure. Thus there is no prospect of help from a distinction like the one made by Mill in *Utilitarianism* (1986, pp.258-262) between higher and lower pleasures, based on the idea of well-informed preferences. However low the pleasure might be, it is still a pleasure and must therefore count in favour of any action that is likely to produce it.

Nor does it seem likely that the difficulty will end there. So long as we attempt to make use of natural goods (those unrestricted by any prior moral assessment) the same objection will recur. Consider, for example, the option of taking the maximising of preference satisfaction to be the aim, whether uniquely or as a part of some more complex objective. Such a theory is bound to leave open the possibility that person A will find that his sadistic impulses give him a strong preference for causing B to suffer. Are we prepared to say that the chance of satisfying such a preference counts morally in favour of the act of inflicting the suffering? Not, surely, if we had already agreed that the pleasure that A would derive fails to count in this way, since a preference deriving from a repugnant pleasure is likely to be no less counter-intuitive in this role than the pleasure itself would have been. The same kind of problem can arise if we substitute for pleasure such goods as happiness or well-being. Provided that we treat these as natural goods, it would seem highly implausible to suppose that repugnant pleasures could never give rise to any happiness, or never contribute to a person's well-being. To rule out such possibilities surely requires instead an appeal to some evaluative conception of the goal, rather as we find, for example, in Scarre's (2001, p.109) elucidation of happiness as the pursuit of 'worthy' objects. Otherwise, I suggest, it is likely that for any candidate for the natural good we will be unable to rule out the possibility of its arising from, or constituting, some condition that will be regarded as morally so deplorable that it is thereby disqualified from providing moral support for an action that produces it.

It would seem, then, that the only solution for consequentialism is one that comes with considerable theoretical costs: the imposition of some kind of evaluative constraint on the entire conception of the good that we are required to promote. As has been argued, partial constraints of this kind will not succeed. Merely to introduce some moralised elements into a complex and partly natural account of the good will be insufficient. Take, most obviously, the long-standing philosophical effort to establish a barrier to victimisation in a consequentialist framework through the inclusion of the upholding of rights among the goals (for example Sen, 1982). In essence this kind of tactic is a moralised version of the rules considered earlier, and thus faces the same objection: we would still have to be able to make sense of the idea that the victim's right not to be tortured was in *moral* competition (albeit victoriously) with the value of the sadistic pleasure of the torturer. Or, to put the point more generally, the likelihood of producing any of the remaining natural goods, however repugnant they might be, would have to be accepted as counting morally in support of abandoning the right (or rule) that prohibited the actions that would generate those natural goods. Avoiding this implication will therefore require constraints of a more comprehensive and radical nature. As it has been expressed by Robert Goodin (1986) in the context of discussing the 'laundering' of repellent preferences in social choice theory, we will need 'input filters' rather than relying on 'output filters' alone; input filters being those that work 'by refusing to count certain classes of desires and preferences when aggregating individual utilities', while output filters work 'by removing certain options from social consideration, whatever their utility' (p.78). Or, put in terms of the foregoing discussion, what is needed is a way of excluding certain pleasures from moral consideration, not (or not only) a way of ensuring that they never contribute to the justification of an action as being morally right.

The precise form of the minimum (assuming that to be a virtue in a theory) constraints required to solve the problem, and the question of whether they are compatible with consequentialism, are issues too large to take up here in detail. They would, after all, call not only for a study of the various possible filtering devices but also a decision on the complex question (see for example Parfit, 1985, pp.24-27, and Griffin, 1992) of exactly what is to count as consequentialism. We can, though, say something briefly about the general kind of constraint that seems to be needed and about its implications for consequentialism in general. Goodin's work is illuminating in this context because, while his concern is with social choice theory rather than moral theory, his problem is a similar one and his aim is the attractively minimalist one of arguing that much useful laundering of preferences may be achieved on grounds 'internal to the preferences themselves' (Goodin, 1986, p.81), so remaining within the sphere of information about utilities. That is, he is considering the extent to which this version of the problem may be resolved prior to the imposition of an evaluative constraint from outside the initial theoretical framework. For our purposes the most directly relevant of his specific suggestions (p.83) is that in the aggregation of preferences the decision-makers may justly ignore certain of an individual's explicit first-order preferences where these are in conflict with the (implied) preference he has for the social situation in which they could be satisfied. One such example, he suggests, occurs with the sadist, the person who has an explicit preference for humiliating others. Given, Goodin argues, that the satisfaction of this preference actually requires that as a public policy the dignity of others be upheld (otherwise the sadist would have no-one worth working on) it can be inferred that the sadist has an implicit but 'logically central' (p.85) preference for this policy. The decision-makers are thus obliged to exclude his sadistic first-order preferences from their construction of a social choice, and to do so on the basis of the sadist's own utilities.

Unfortunately it is clear, I think, that we cannot solve our problem in moral theory simply by transferring this minimal filter from the theory of social choice. All that it could give us, presumably, is the argument that a person's preference for (or pleasure in) sadistic acts is inevitably disqualified as a moral reason for a social rule in favour of sadism, on the grounds that the establishment of the rule would be illogically self-defeating of that person's preference. Such an argument would do nothing to exclude the preference (or the pleasure) as a moral reason for individual action in the context of the social maintenance of human dignity. More generally, the limited moral effect of this kind of preference-based filter is evident if we note that the reason given by the theory for disqualifying the sadistic preference from any influence on social rule-formation is not the repugnance of the preference but its inconsistency with a preference for the rule. Indeed, it would seem that not only the effect but the very point of the filter is actually to rescue the individual's sadistic preference from inevitable frustration.

Suppose then that we cross the boundary into evaluation and turn to a constraint that more directly reflects our actual moral concerns, such as the kind found in the consequentialist theory of G.E. Moore. In Moore's work, at least at the time of *Principia Ethica*, there is clearly in operation an evaluative restriction on the sort of pleasure that should be regarded as good, since he asserts that the enjoyment of pain suffered by others is an evil state, the evil of which would actually be heightened by any increase in the enjoyment (1903, p.210). To examine the implications for consequentialism of this general kind of restriction let us consider (while leaving open whether it has any merit in itself) an explicit but relatively simple principle in the same spirit as Moore's assertion: the right act is that which will probably maximise the total of morally acceptable pleasure. There is a good chance that such a principle will exclude non-consensual sadistic pleasures from any counter-intuitive place in our moral explanations, but only at the obvious cost of adding some substantive and controversial moral input prior to the consequentialist element of the theory. Plainly, for the many possible principles in a similar vein, one could only make individual assessments of the likelihood of successful exclusion and of the various costs to the moral theory. For example, the 'objective list' theory favoured by David Brink (1989, p.264) seems to have the exclusion of sadistic pleasures and preferences among its explicit objectives, to be achieved through the presence of moral constraints (including respect for others) on what is to count as a valuable project. It thus gives us something more comprehensive by way of exclusion than the simple rule mentioned previously. However, the costs may be great, perhaps going beyond the evident problems of adding greater complexity to our judgement of what is good to the point of threatening the operational element of the theory. That is, as the concept of the good is developed along overtly moral lines it becomes more likely that the principle of its being maximised or promoted will be challenged. As Griffin points out in his discussion of broad and narrow consequentialism (1996, pp.161-166), once our conception includes such purely moral goods as acts of justice, honesty, or the avoidance of lies, the very idea of promoting the good is liable to seem morally less well motivated, since it will amount to 'the promotion of good behaviour in humanity at large' (p.164).

IV

What, finally, does all this imply with respect to the long-standing debate over the theoretical ordering of the right and the good? The argument of this article has been that, as moral theories, the various forms of utilitarianism (and consequentialism more generally) must not only identify right actions but invoke only morally plausible explanations in so doing, and that this requires that any appeal to the generation of certain dubious kinds of goods be excluded as morally counter-intuitive. The effect of such a requirement, it has been suggested, is to necessitate a constraint on the conception of the good that will, in some form or other, restrict us to promoting solely those goods whose production is held to be morally acceptable. Does this constitute the priority of the right over the good? This seems to be suggested tentatively by Griffin (1986, p.321) when he considers whether moral theory might need a narrow conception of well-being, where 'the sordid side of human nature' does not appear. Yet it is not clear that this kind of constraint requires that we must in any straightforward way define the right prior to the good. For example, it does not show that we only believe sadistic pleasures to be bad because we believe sadistic acts to be wrong, since the badness of pleasure felt at the suffering of others might itself be fundamental to our moral thinking. What it does show, perhaps, is that while the conception of the good in a moral theory might be basic, it is an essentially moral conception. And thus, given the nature of people as we know and (sometimes) love them, we would have to reject another of Griffin's (1986, p.72) tentative conclusions: that the conception of well-being that we develop in prudential value theory will serve for morality also.

REFERENCES

Beauchamp, T.L. and Childress, J.F. (2001) *Principles of Biomedical Ethics*, fifth edition (Oxford: Oxford University Press).
Bentham, J. (1967) *A Fragment on Government and An Introduction to the Principles of Morals and Legislation*, ed. W. Harrison (Oxford: Basil Blackwell).
Brink, David O. (1989) *Moral Realism and the Foundations of Ethics* (Cambridge: Cambridge University Press).
Goodin, Robert E. (1986) 'Laundering Preferences', in J. Elster and A. Hylland (eds.) *Foundations of Social Choice Theory* (Cambridge: Cambridge University Press).
Griffin, James (1986) *Well-Being: Its Meaning, Measurement and Moral Importance* (Oxford: Clarendon Press).
----- (1992) 'The Human Good and the Ambitions of Consequentialism', in E.F. Paul, F.D. Miller and J. Paul (eds.), *The Good Life and the Human Good* (Cambridge: Cambridge University Press).
----- (1996) *Value Judgement* (Oxford: Clarendon Press).
Hare, R.M. (1981) *Moral Thinking* (Oxford: Clarendon Press).
Hooker, Brad (2000) *Ideal Code, Real World: A Rule-Consequentialist Theory of Morality* (Oxford: Clarendon Press).
Kagan, Shelly (1989) *The Limits of Morality* (Oxford: Clarendon Press).
Mill, J.S. (1986) *Utilitarianism*, ed. Mary Warnock (Glasgow: Fontana Press).
Moore, G.E. (1903) *Principia Ethica* (Cambridge: Cambridge University Press).
Nussbaum, Martha (2000) 'Why Practice Needs Ethical Theory: Particularism, Principle, and Bad Behaviour', in B. Hooker and M. Little (eds.) *Moral Particularism* (Oxford: Clarendon Press).
Parfit, D. (1985) *Reasons and Persons* (Oxford: Oxford University Press).
Rawls, John (1999) *A Theory of Justice*, revised edition (Oxford: Oxford University Press).
Scarre, Geoffrey (1996) *Utilitarianism* (London: Routledge).
----- (2001) 'Upton on Evil Pleasures', *Utilitas* 13, 106-111.
Sen, Amartya (1982) 'Rights and Agency', *Philosophy and Public Affairs* 11, 3-38.
Smart, J.J.C. (1973) 'An outline of a system of utilitarian ethics', in J.J.C. Smart and Bernard Williams, *Utilitarianism: For and Against* (Cambridge: Cambridge University Press).

Smith, Michael (2000) 'Does the Evaluative Supervene on the Natural?', in R. Crisp and B. Hooker (eds.) *Well-Being and Morality: Essays in Honour of James Griffin* (Oxford: Clarendon Press).

Sprigge, T.L.S. (1988) *The Rational Foundations of Ethics* (London: Routledge & Kegan Paul).

Upton, Hugh (2000) 'Scarre on Evil Pleasures', *Utilitas* 12, 97-102.

INDEX

A

academic progress, 120
academics, 120, 122, 123, 124, 125
acceptance, 45, 134, 139
access, 120
accountability, 14
accounting, 7, 103
accreditation, 114
achievement, 116
acid, 74
activation, 67, 68, 69, 70, 71, 73, 74, 75, 80, 81, 85, 89, 90, 92
adaptation, 22
adsorption, 82
adulthood, ix, 113, 118, 120, 121
adults, 120, 121, 123, 125
advertising, 27
affect, 45, 57, 58, 61, 68, 99, 121
Africa, 63, 78
age, viii, 21, 22, 25, 26, 31, 35, 123
agent, 29, 53, 137
aggregation, 70, 141
agriculture, vii, 86
AIDS, 78
alcohol, 13, 36
alternative, ix, 27, 35, 43, 47, 66, 67, 73, 74, 75, 77, 97, 98, 120
alternatives, 47
amphibia, 90
amyotrophic lateral sclerosis, 65, 78
animals, 30, 66, 72, 73, 77
anthropology, 34
antibody, 65, 69, 70, 75, 79, 81, 82, 84, 85
anticoagulant, 68
anxiety, 28, 132
argument, 4, 47, 49, 50, 54, 60, 77, 88, 98, 102, 103, 136, 142, 143
Aristotle, 54
assault, 39
assessment, 13, 122, 123, 125, 131, 134, 137, 140
assets, 17
association, 12, 57, 61
assumptions, viii, 39, 41, 57, 60, 61, 126, 130
asymmetry, 90, 94
attachment, 67, 69, 75, 85
attention, 3, 18, 51, 98, 118, 126, 133, 137
attitudes, 42, 116
attribution, viii, 39, 41
Australia, vii, 1, 10, 11, 12, 14, 17, 18
authority, 6, 7, 8, 10, 11, 30, 52
autonomy, viii, 39, 40, 41, 43, 44, 45, 47, 50, 51, 52, 53, 56, 58, 60, 61, 100, 115, 122
avoidance, 133, 139, 143
awareness, 11, 14, 116

B

back pain, 57

bacteria, 80
banks, 60
barriers, viii, 63, 66
bias, 90, 133
binding, 44, 75, 80, 85
biosafety, 66
birth, 33
bleeding, 46
blood, 72, 83
body, 25, 27, 30, 36, 45, 77, 91, 114
bonds, 17
bone marrow, 65, 78
brain, 65, 76, 78
breaches, 45
business organisation, 121

C

campaigns, 27
Canada, 124
cancer, 65
cancer cells, 65
cast, 124
Catholic Church, 25
causation, 58
CD8+, 70
cell, 64, 65, 68, 70, 71, 72, 78, 79, 80, 81, 84, 88, 89, 90, 91, 92, 94, 95, 96
cell fate, 90, 96
cell surface, 79
chain of command, 3, 7, 8, 11, 18
chemotaxis, 76
childhood, 67
children, 121
chimpanzee, 72
cholesterol, 71
Christianity, 25
chromosome, 89
circulation, 72
classes, 141
classification, 5, 35
classroom, 119, 121, 126, 127
cleavage, 75, 90, 91, 95
clinical trials, 37, 65
cloning, 35, 88, 94, 95

clusters, 73, 90
coagulation, 68, 69
cognitive dissonance, 123
collaboration, 115, 117
colleges, ix, 113
coma, 27
commitment, ix, 5, 9, 44, 103, 104, 113, 114, 115, 117, 119, 123, 124, 127
commodity, x, 129, 130
communication, 2, 13, 42, 43, 44, 48, 91
community, 15, 17, 23, 26, 33, 43, 50, 51, 54, 57, 58, 59, 60, 77
compensation, 41
competence, 53
competition, 141
complement, ix, 18, 64, 67, 68, 69, 70, 71, 73, 74, 75, 76, 78, 80, 81, 83, 84, 85
complexity, 120, 143
compliance, 3, 8, 9, 13, 44, 45
components, 11, 68, 75, 85, 107
composition, 31
comprehension, 27, 29
compulsory education, 120
concentration, 29, 89
conception, x, 30, 33, 36, 43, 44, 51, 53, 58, 100, 129, 130, 131, 138, 139, 140, 143
conduct, 3, 7, 8, 9, 11, 13, 19, 116, 118, 123, 125, 137
confidentiality, 120, 121
conflict, 141
confusion, 37
consciousness, 29, 33, 35
consensus, vii, 2
consent, viii, 4, 39, 40, 41, 42, 43, 44, 45, 46, 47, 48, 49, 50, 51, 53, 54, 55, 57, 58, 60, 61, 130
consequentialism, 133, 137, 140, 141, 142, 143
conspiracy, 7, 26, 28, 137
constitution, 36
construction, 43, 89, 134, 142
consumption, 77
context, 2, 16, 31, 33, 35, 40, 43, 44, 45, 50, 54, 59, 117, 119, 122, 123, 126, 130, 141, 142

control, ix, 7, 8, 16, 25, 45, 46, 61, 64, 73, 74, 80, 84, 85, 88
conviction, 7
corporate life, 3, 8, 116
corporate scandals, 7
corporations, 3, 7
corruption, 12, 13
corticosteroids, 72
costs, 53, 56, 59, 140, 142
coverage, 11
creative teaching, 124
credibility, 122
credit, 16, 135
crime, 130, 135, 136
criminals, 137
critical thinking, 119
criticism, 41, 98, 107, 121, 130, 136
crystals, 35, 36
culture, vii, 1, 2, 3, 6, 7, 8, 9, 14, 15, 18, 19, 24, 25, 29
curiosity, 124
curriculum, 114
customers, 15, 24, 35
cyclosporine, 72
cytokines, 70, 75
cytotoxicity, 70

D

damage, 46, 57
danger, 17, 37, 60
death, 27, 28, 33, 34, 65, 76, 89, 93
decay, 73, 83, 84
decision making, 3, 13
decisions, 4, 32, 36, 52, 53, 57, 96, 121, 122, 133
defendants, 40
defense, 72
deficit, 57
definition, 11, 22, 32
delivery, 78
demand, 5, 27, 76, 119
dendritic cell, 70
denial, 132
Denmark, 97

deontology, 2
deposition, 69
desire, 23, 122, 123, 138
destruction, 70
detection, 16
deterrence, 136
diabetes, 78
differentiation, 64, 71, 89, 90
diffusion, 26
dignity, 141
diploid, 88, 92
directives, 30
disability, 56
discipline, 118
disclosure, 48, 49
discourse, 44, 49
discrimination, 13
disorder, 65
disposition, 22
distribution, 47, 54, 55, 57, 59, 100, 101, 105, 106
distributive justice, 4, 49, 54, 55, 56, 57
disutility, 136
diversity, 14
division, 18, 90, 91, 95
DNA, 89, 92, 95
doctors, 41, 47, 59
donors, 66, 83
drugs, 12, 13, 72
due process, 4, 18

E

ecology, 11
edema, 69
educational process, 118, 120, 122, 123, 125
egalitarianism, ix, 97, 98, 99, 100, 101, 102, 103, 104, 106, 107, 109, 110
egg, 91, 92, 95
embryo, ix, 87, 88, 90, 91, 93, 94, 95, 96
embryogenesis, 95
embryology, 88, 94
emotion, 31
emotions, 6, 31
empathy, 13

employees, 11
employment, 116
empowerment, 115
encephalitis, 77
endothelial cells, 67, 68, 69, 70, 80, 81, 85
endothelium, 67, 69, 83
England, 56, 59
environment, vii, 3, 6, 8, 11, 89, 92, 124
enzymes, 71
epigenetics, 94, 96
equality, ix, 3, 7, 8, 97, 98, 99, 100, 101, 102, 104, 105, 106, 107, 109, 115, 117
equilibrium, 37, 138, 139
equity, 4
erythropoietin, 71
ethical issues, 5, 14, 16, 77
ethical standards, 114
ethics, 2, 8, 9, 13, 14, 15, 16, 18, 19, 94, 117, 118, 137
Europe, 26
euthanasia, 34
evidence, 13, 36, 40, 78, 80, 93, 116
evil, 22, 28, 29, 135, 142
evolution, 31
examinations, 122
exclusion, 22, 137, 142
exercise, 42, 125
expectation, 15, 115
experts, 88
expression, 23, 24, 31, 73, 79, 83, 84, 88, 89, 92, 119, 120, 136

F

failure, 11, 23, 41, 56, 58, 60, 68, 82, 92, 137
fairness, 3, 4, 7, 8, 13, 17, 122, 123, 125, 127
faith, 28, 76
false belief, 136
family, vii, 2, 10, 15, 16, 18, 27, 56, 77
fat, 28
fauna, 11
fear, 11, 77, 121
feedback, 123
feelings, 6, 130, 135

fertilization, ix, 33, 34, 87, 90, 91, 92, 93, 94, 95
fibrin, 69
fibroblasts, 92
fish, 28
flora, 11, 80
fluctuations, 89
fluid, 69, 73
focusing, 43
food, vii, 55, 77
food production, vii
fraud, 7, 16
freedom, ix, 31, 54, 113, 118, 119, 121, 125
fuel, 59
funding, 49
furniture, 7

G

gastrointestinal tract, 67
gastrulation, 90, 93
gene, 65, 67, 73, 89, 95
gene expression, 89, 95
gene therapy, 73
generation, 33, 73, 83, 130, 143
genes, 88, 89, 92
genetic information, 88, 90
genome, 88, 90, 96
genre, 31
germ layer, 90
gestation, 66
glycosylation, 79
goals, 3, 5, 7, 8, 9, 48, 53, 140
God, 22, 27, 35
goods and services, 5
governance, 96
government, 16, 17, 18, 115, 117, 124
grades, 119, 121
grading, 81, 124
grants, 46
granules, 76
Great Britain, 31
grouping, 6
groups, vii, 2, 26, 35, 50, 74, 77
growth, 5, 89

guidance, 37
guidelines, 14, 15
guilt, 33, 131
guilty, 13, 130
Guinea, 82

H

habitat, 11
happiness, 5, 23, 25, 28, 130, 133, 135, 136, 140
harm, 3, 31, 53, 56, 58, 60, 125, 133, 136
harmony, 3, 31, 125
health, 53, 95
heart transplantation, 82
heart valves, 65
hedonism, 134
hemorrhage, 74
higher education, ix, 113, 114, 115, 116, 117, 118, 119, 120, 121, 122, 123, 124, 125
hip, 13
histone, 89
hormone, 71
host, 69
hub, 22, 29
human development, 93
human dignity, 95, 142
human rights, 4
human xenotransplantation, ix, 64
humility, 7
humus, 22
hybrid, 88
hypothesis, 36

I

ideas, 6, 7, 28, 34
identification, x, 114, 116, 139
identity, ix, 51, 61, 90, 113, 118, 127
illumination, 35
imagination, 25
imbalances, 58
immune response, 74
immune system, 65, 84
immunoglobulin, 80
immunosuppression, 83
immunosuppressive drugs, 71, 72, 77
implementation, 105, 115
imprinting, 89
in vitro, 70, 73, 75, 91
inclusion, 140
incongruity, 130
independence, 50, 120
indication, 42
individual action, 142
individual students, 122
individualism, 24
induction, 82
industry, 11, 120
inequality, 4, 57, 101, 102, 105, 106, 107, 123
infinite, 22, 31
inflammatory cells, 70, 74
influence, 25, 26, 27, 32, 51, 60, 142
informed consent, 41, 42, 46, 47, 48
inheritance, 96
inhibition, 84
inhibitor, 64, 69, 73, 83
initiation, 25
injury, 41, 59, 60, 83, 84
innovation, 17
input, 141, 142
insight, 137, 138
inspiration, 28, 31
instinct, 23
institutions, 114, 119, 124
instruction, 126
instructors, 12, 120
instruments, 27, 31, 35
insurance, 60
integration, ix, 87
integrity, 3, 5, 8, 9, 13, 14, 15, 17, 46, 80, 127
intellect, 22, 23, 28
intellectual property, 16
intelligence, 5
interaction, 57, 67, 84, 91, 95
interactions, 54, 55
interest, 3, 7, 10, 12, 13, 16, 35, 98, 104, 121, 135
interference, 49, 61

interpretation, 25, 29, 34, 37, 50, 104
intervention, 45, 49, 54, 55
intrinsic value, 100, 104, 106, 109, 138
intuition, 28, 99
isotonic solution, 72
Italy, 21, 32, 35

J

Japan, 77
justice, 4, 7, 8, 16, 47, 54, 55, 57, 58
justification, viii, 34, 39, 41, 42, 50, 53, 58,
    109, 131, 137, 141

K

kidney, 71, 82
kidneys, 71, 72, 83
killer cells, 64
kinetics, 71
knowledge, vii, 2, 22, 25, 26, 27, 28, 30, 37,
    57, 90, 114, 118, 119, 138

L

language, 22
lead, 23, 26, 27, 29, 31, 37, 88, 124, 135, 138
leadership, 3, 6, 17, 18
learner support, 122
learners, ix, 113, 115, 116, 117, 118, 119,
    120, 121, 123
learning, 114, 115, 116, 117, 119, 121, 123,
    125, 126
learning environment, 119
learning outcomes, 114
legislation, 51
liability, 55, 58
Libertarian, 49, 51, 54, 55, 56, 58, 61
life experiences, 27, 121
likelihood, 58, 131, 141, 142
listening, 32
litigation, 59
liver, 71, 72, 82
liver transplantation, 82

location, 89
logistics, 77
love, 143
low risk, 59
loyalty, 17
lymphocytes, 81
lysis, 73

M

macrophages, 70, 72
Malaysia, 77
management, 3, 7, 8, 9, 15, 16, 17, 18, 119
manipulation, 94
manufacturing, 11
market, 11, 24, 120
marriage, 93
marrow, 72
mass, vii, 27, 91, 123
mass media, vii
mast cells, 76, 85
maturation, 71
MCP, 64, 73
measures, 100
media, 11, 12, 27
memory, 89, 91
men, 18, 28, 33, 35
mental health, 48
mental state, 42, 136, 139, 140
mental states, 140
methodology, 36, 37, 130, 138
methylation, 89, 90, 92, 95
Mexico, 32
mice, 72, 73, 79, 84, 96
military, 16
mimicry, 85
mining, 11
minority, 58
MIP, 64, 75
mixing, 96
mobility, 65
mode, 55
models, 41, 47, 72, 74, 75
molecular immunobiology, 71
molecules, 68, 69, 80, 92

money, 12, 55, 59, 121
monoclonal antibody, 81
moral code, 3, 7
moral reasoning, 5
morality, 7, 45, 52, 143
mosaic, 55
motivation, 13, 116
movement, viii, 21, 22, 26, 31, 89
music, 30, 31, 32
musicians, 31
mutual respect, 10, 119
mutuality, 44

N

natural killer cell, 76, 81
necrosis, 64, 69
needs, 3, 7, 35, 43, 47, 56, 70, 118, 119, 122, 124, 136
negotiation, 48, 49, 51, 54
nerve, 57, 65, 127
network, 89
neutrophils, 70
NK cells, 70
normal development, 91
novelty, ix, 87
nuclei, 91, 92
nucleus, 88, 95
nursing, 48

O

obligation, 14, 47, 49
observations, 81
obstruction, 7
oocyte, 88, 90, 92
openness, 13
organ, 65, 66, 69, 73, 76, 81, 83, 85
organism, 89, 90, 91, 92, 93
organizations, 11, 121
orientation, 91
outline, 144
output, 5, 141
ownership, 45

P

pain, 135, 142
paralysis, 57
parental involvement, 120
parents, 88, 116, 120, 121
Parkinson's disease, 65, 78
parthenogenesis, 88
passive, 45
pathogenesis, 80
pathways, 74
peers, 12, 121
peptides, 74, 75
perfusion, 83
permit, 49, 50
personal autonomy, 53
personal relationship, 123
personhood, 34
perspective, 49, 93, 119, 122
Perth, 10, 11
phage, 74
phenotype, 89, 91, 92, 93
pigs, 66, 71, 73, 77, 83, 84
placenta, 91
planets, 25
plants, 30
plasma, 72
plasminogen, 64, 69
plasticity, 93
platelets, 69
plausibility, 134, 136, 139
pleasure, x, 129, 130, 131, 132, 134, 135, 136, 138, 139, 140, 141, 142, 143
polar body, 91
polarity, 90, 93
police, 8, 9, 11, 12, 13, 15, 18
politics, 96
polypeptide, 84
poor, 16
population, 52
power, 3, 7, 8, 29, 30, 33, 47, 57, 58, 61, 119, 120, 121, 122, 123, 125, 137
pragmatism, 134
predicate, 103
preference, 130, 140, 141, 142

pregnancy, 56
preparation, 24, 123
pressure, 124
prevention, 33
primate, viii, 63, 70, 72, 79, 83
principle, viii, ix, 3, 7, 11, 29, 33, 37, 54, 56, 58, 61, 63, 99, 100, 101, 102, 103, 105, 110, 113, 118, 119, 134, 135, 136, 137, 142
prisoners, 132
privacy, 4, 121
probability, 120
production, x, 69, 95, 129, 130, 132, 138, 143
professional duties, 15
professionalism, 117
program, 11, 12, 89, 92
proliferation, 89
proposition, 42
proteins, 68, 73, 74, 83, 89
proteoglycans, 68, 76
proteolysis, 80
psoriasis, 79
psychology, 52
public opinion, 35
public policy, 141
punishment, 34, 130, 135, 136
pupil, 118
pure water, 36

R

range, 2, 12, 13, 14, 16, 18, 46, 51, 59, 60, 78, 131
ratings, 124, 126
rationality, 52, 53
reading, 28
reagents, 77
reality, 28, 35, 37, 48, 73, 90
reasoning, 108, 118
recall, 31, 135
receptors, 71
recognition, 51, 52, 53, 81, 114, 123, 124
reconcile, 25, 96
recovery, 22, 56, 59
reduction, 72, 84
redundancy, 121

reflection, 32, 52, 88, 91, 114, 115, 117, 118
reflective practice, ix, 113, 117
regression, 36
regulation, 61, 74, 89, 91
regulators, 67
relationship, viii, 16, 27, 39, 41, 42, 44, 48, 49, 51, 60, 120, 121, 123, 125, 127
relationships, 57, 123, 125
relevance, 51, 56, 93, 133, 135
religion, vii, viii, 21, 22, 25, 28, 30
religiosity, 28
reputation, 17
research funding, 124
resistance, 29
resolution, 59
resources, 16, 47, 49, 54, 55
responsibility, viii, 3, 10, 11, 14, 18, 39, 41, 47, 53, 54, 55, 57, 58, 59, 60, 61, 94, 116, 120, 121, 122
restructuring, 127
retention, 76
rewards, 9
rights, 4, 11, 16, 18, 45, 49, 60, 61, 77, 116, 137, 140
risk, 41, 56, 57, 58, 77, 86, 121
rugby, 8

S

sacrifice, 23, 29, 34
sadism, 131, 132, 142
sales, 7
satisfaction, 131, 140, 141
scholarship, ix, 113, 114, 115, 117, 123
school, ix, 113, 116, 119, 120, 122
schooling, 118
scientific knowledge, 37
scientific progress, 36, 77
search, 35
Second World, 40
secretion, 69, 70
securities, 7
self, 10, 13, 29, 32, 36, 40, 43, 45, 50, 51, 53, 58, 89, 90, 93, 119, 121, 133, 142
self-identity, 45

self-organization, 90, 93
sensations, 24, 31
series, ix, 16, 31, 56, 89, 113, 118
serum, 71
services, 9, 10, 12, 15
severity, 70
shame, 135
shape, 61, 95
shaping, 89
shares, 100, 101, 102, 106
sharing, 31, 60
shortage, 65, 76, 85, 133
sign, 25
signals, 92
Singapore, 77
skills, 119
skin, 65
social context, 53
social policy, 109
social relations, 51
social responsibility, 11
social structure, 6
somatic cell, 88, 92
South Africa, 65
Spain, 87
span of control, 6
specialisation, 6
species, viii, 63, 66, 71, 73, 82, 84
specificity, 100
spectrum, 65
speech, 4
sperm, 91, 92, 95
spirituality, 24, 31
stability, 100
stages, ix, 87, 118, 122
stakeholders, 120
standards, 6, 9, 15, 19, 115, 116, 124, 125, 126
Star Wars, 29
stars, 29
steel, 40
sterilisation, 56
stigma, 77
stimulus, 88
strategies, viii, 7, 12, 14, 18, 64, 71, 72

streams, 14
strength, 11, 17, 61
stress, 56, 100, 130
structural changes, 89
students, ix, 113, 114, 115, 118, 119, 120, 121, 122, 123, 124, 125, 126, 127
subsistence, 27
substitution, 140
sugar, 73
suicide, 34, 65
Sun, 84
supervision, 14, 123
supervisor, 13
supply, 11, 76
suppression, 33
surplus, 95
surveillance, 75
survival, viii, 63, 65, 71, 72, 73, 78, 81, 84
Switzerland, 92
symmetry, 90
synchronization, 92
synthesis, 71, 73
systems, 120, 122

T

teachers, ix, 113, 114, 115, 117, 118, 119, 120, 121, 122, 123, 124, 125, 127
teaching, ix, 113, 114, 115, 116, 117, 118, 119, 121, 123, 124, 125, 126, 127
technology, 5, 6, 15, 65
telephone, 13, 16
tension, 22, 121, 122
T-helper cell, 70
theory, viii, 2, 3, 4, 5, 7, 18, 21, 22, 23, 25, 33, 36, 37, 48, 55, 109, 121, 131, 133, 135, 136, 137, 138, 140, 141, 142, 143
therapeutic approaches, 74
therapeutic process, 37
therapeutics, 84
therapy, 35, 36, 81
thinking, 7, 17, 138, 143
threat, 17, 59, 137
threats, 120
thrombomodulin, 68, 69

time, viii, 3, 7, 8, 12, 16, 17, 21, 22, 24, 27, 28, 30, 31, 40, 47, 48, 68, 69, 89, 91, 93, 115, 119, 120, 123, 124, 133, 142
tissue, viii, 63, 65, 69
tissue plasminogen activator, 69
TNF, 64, 75
torture, 132, 134
toys, 27
TPA, 64, 69
trading, 7
tradition, 25, 118, 124
training, vii, 2, 8, 9, 12, 13, 19, 29
training programs, vii, 2, 8, 12
transactions, 12, 55
transcription, 69
transfection, 73
transformation, 92, 95
transgression, 58
transition, 120
translation, 22, 26, 28, 29, 31, 35, 36
transmission, 33
transparency, 2, 125
transplantation, viii, 63, 65, 66, 67, 69, 71, 72, 73, 76, 78
trend, 23
trial, 12
trust, 17, 37, 133
tuition, 120
tumors, 65, 78
tutoring, 123

U

UK, 92, 113, 114, 115, 116, 117, 118, 120, 123, 126, 127, 129
uniform, 93
United States, 7, 64, 88
universe, 27, 136
universities, 118, 121
uric acid, 71

V

values, vii, ix, 1, 2, 6, 13, 15, 16, 17, 18, 44, 51, 52, 113, 114, 116, 117, 119, 124, 125
variable, 51, 52, 57
variables, 51
variance, 135, 139
variation, 89
vector, 28, 29, 73, 78
victimisation, 140
victims, 131, 134
viral vectors, 65
virology, 85
virus infection, 86
viruses, 77, 86
vision, 22, 26, 27, 29
voice, 30, 120
voluntarism, viii, 21, 24

W

Wales, 129
water, 36, 41, 60
welfare, 28, 56, 57, 60, 98, 99, 100, 101, 102, 104, 105, 106, 107, 108
well-being, 133, 140, 143
western culture, 24
witnesses, 12
words, 11, 15, 18, 23, 29, 33, 34, 36, 37, 43, 55, 105, 106, 107, 109, 124, 131, 132, 135, 137
work, 3, 6, 8, 14, 17, 18, 27, 32, 54, 77, 90, 114, 119, 125, 132, 136, 137, 141, 142
workers, 6
workplace, 4
World Health Organization, 110
writing, 115
wrongdoing, 10, 19, 54, 57

X

xenografts, viii, 63, 66, 70, 71, 72, 73, 77, 79, 81, 83, 84 xenotransplantation, viii, 63, 65, 66, 70, 71, 72, 73, 77, 78, 81, 82, 83, 86

Z

zero sum game, 124
zygote, 88, 89, 90, 91, 92, 93